# Mentoring in Schools

# Mentoring in Schools: A Handbook of Good Practice

SARAH FLETCHER

Routledge
Taylor & Francis Group
LONDON AND NEW YORK

First published in 2000 by Routledge



Routledge

2 Park Square, Milton Park, Abingdon, Oxfordshire OX14 4RN

711 Third Avenue, New York, NY 10017

First issued in hardback 2016

*Routledge is an imprint of the Taylor and Francis Group, an informa business*



**British Library Cataloguing in Publication Data**

A CIP record for this book is available from the British Library.

Typeset by JS Typesetting, Wellingborough, Northamptonshire

ISBN 13: 978-1-138-15747-7 (hbk)
ISBN 13: 978-0-7494-3183-9 (pbk)

I dedicate this book to the memory of my father,
Robert Patchin,
who was my finest mentor.

# Contents

## Part 5. Developing competence through mentoring

# Foreword

This book is published at a time of national and international concern to enhance the professional standing of teachers. Sarah Fletcher's passion and commitment is to help to enhance professionalism in education through mentoring. She believes that the mentoring relationship is central to the process of improving learning for both pupils and teachers. In a carefully structured text she moves her readers through the values, skills and understandings that are vital in enhancing the quality of mentoring in teacher education.

The contents of this book are informed by her experiences as a secondary school teacher and mentor for some 20 years, as well as by her university teaching and mentoring since 1994 in the Department of Education of Bath University. She is one of the growing numbers of university teachers who are researching their own practice in order to understand how to help their students to improve the quality of their learning.

In this contribution to the development of mentoring in local, national and international communities of mentors and teacher researchers, Sarah Fletcher demonstrates a sensitive awareness of the tensions that can arise between institutional structures and the values that motivate individual mentors. She understands the importance of affirming the importance of educational values within communities of mentors and of the need to develop supportive structures for enhancing the quality of mentoring. Most importantly, she shows that she understands the importance of teachers telling their own stories as they sustain and develop their professional identities in times of rapid change.

Her mentoring Web site (www.mentorresearch.net) provides the opportunity for you to make this book a part of a living process of improving professionalism in education. You could do this by responding to its contents and contributing your own stories to our growing community of mentors. This book brings together theory and practice through a focus on the importance of relating mentoring to the process of improving the quality of learning for both oneself and one's students.

*Dr Jack Whitehead*
*University of Bath*
*May 2000*

# Preface

How shall I define 'mentoring' for you? Yamamoto's (1990) definition was 'to see life grow'. For me, it is 'to participate in life growing.'

Mentoring has no one sense for as soon as it *is* it has changed. In this book I attempt to share with you what mentoring has come to mean to me as I have mentored in schools. I also draw from my experience of being mentored in an institution of higher education and from working with many mentors.

This is the book that I knew I needed when I undertook mentoring. Of course it reflects the current situation in initial teacher education with its emphasis on meeting TTA standards for qualified teacher status rather than those for the Licensed Teacher Scheme in Bedfordshire where I mentored. Consider this book as *your* mentor. Seek guidance from it thoughtfully but realize that real development in terms of professional expertise cannot come from cloning but comes from interaction between its pages and you. It is less a cover-to-cover commitment than a dip-in-and-read-as-you-need-it guide.

Whether you are a seasoned mentor, looking to develop your skills, or a novice mentor, this book is intended to be more than just a compendium of practice. It contains information on generic issues in mentoring and, although other chapters may seem less applicable to your mentoring, they may well contain appropriate advice.

Perhaps the best way to make use of this book is to prioritize according to your needs. Divide the chapters into three categories thus:

- category one, directly relevant – I need this information now;
- category two, indirectly relevant – I am likely to need this soon;
- category three, looks interesting – and I might need it one day.

Read each of the category one chapters and after each review your own mentoring practice in the light of what you have been reading. This does not mean that you necessarily need to make written notes but you may well find it is useful if you do. Come back to your notes and the relevant chapter later should further questions arise. If you do have questions specific to your mentoring situation, first consider asking a mentor in the same department or in the same school as you so that you can share their thinking. If this is not appropriate, a member of the senior management team in your school is likely to be able to give you an informative view. You could ask a mentor from another school for assistance. If you work in an ITT Partnership seek help and advice from tutors. Remember to keep phone numbers and e-mail addresses to hand!

Above all this is intended as a reference book but it does not and can not supply all the solutions and guidance you are likely to need. There are other excellent mentoring guides, although none covers the same ground in quite the way this one does. In a sense this is an interactive reference book. Mentoring is largely a matter of common sense but this book aims to throw light on areas that lie beyond perspectives of classroom teaching as well as drawing from them. Learn from your own reactions as you learn from the information printed on the page. This book is intended to be a springboard for your thoughts. It is intended to be interesting and to stimulate imaginative and creative mentoring in you!

Whatever brings you to read this handbook, whether you are already a mentor or a school-based or university-based tutor working with mentors, draw up an action plan so you can implement what you are learning. If you are not yet a mentor think about the values and skills you can bring to mentoring and start working *now*. Think of yourself as a professional educator and seek ways to develop your practice, for mentoring enriches teaching as teaching enriches mentoring.

Finally, I offer this book as a starting point in disseminating some of the excellent practice that I observe in working closely with mentors in a number of schools. There are far more examples of 'good practice' than a single handbook can ever encompass and there are also examples of 'good practice' that will remain valid only in a particular context. Be prepared to adapt and monitor as a professional educator – there are no quick fixes and all-embracing solutions in mentoring but there are signposts. The illustrations I offer in the text are drawn from my work.

As a teacher you bring a wealth of skills and understanding with you into mentoring. You can appreciate already that learners need space to experiment, to make mistakes and to be accepted non-judgementally primarily as individuals as well as potential practitioners. Be open to learning as you educate! Revisit this book as your mentoring develops in tandem with the development of your mentee. Revisit it each time you undertake a new mentoring role. Most importantly, use the impetus of this book as a starting point for seeking out and disseminating further examples of 'good practice in mentoring'. You have the potential to know more than I.

Mentoring is something that is done with a trainee and not to a trainee. As you mentor, you will learn and you will teach – and so will your mentee. Mentoring is a dynamic process of interaction and, although this book concentrates on guiding you as you mentor trainee teachers, its principles apply to mentoring in a wider context. Mentoring is concerned with easing transitions and enabling change and growth. This is a process of active, creative engagement in education and not just finding evidence to support externally imposed standards of professional conduct. Mentoring lies at the heart of effective learning schools and brings the TTA's Framework for Professional Development to life. Whether you are a beginning teacher, an aspiring head of department or special needs co-ordinator, or an aspiring headteacher mentoring can assist you.

This book is largely concerned with mentoring with pre-service and newly qualified teachers and I aware that there needs to be a sequel on mentoring for whole-school development, One handbook of good practice cannot do justice to the potential that mentoring holds for developing and creating professional knowledge but I hope it provides at least a suitable starting point.

I would like to express my thanks to my family and colleagues for their support and for the information that appears as illustrations of mentoring throughout the text. In particular I would like to thank John Hewitt, Jack Whitehead and Jen Russ for mentoring me in my writing.

Thanks are also due to Mike Berrill, Mike Bosher, Trish Shubrook, Kate Brockbank and Gordon Trafford for arranging for me to undertake research in their schools: Challney High School for Boys, Luton; St John's School, Marlborough; The Clarendon School, Frome; Cirencester Deer Park School, Cirencester and The John Bentley School, Calne.

# Introduction: what is mentoring?

In this chapter we shall be considering the following:

- A definition of mentoring.
- How this book is organized.
- Mentoring within initial teacher training.
- Identifying good practice in mentoring.

## A definition of mentoring

'Mentoring' is the term, now in wide use within many professions, that reflects the potential of a one-to-one professional relationship that can simultaneously empower and enhance practice. In business, it is regarded as distinct from coaching (which usually indicates intensive focusing to improve a particular skill) and assessment. However, 'mentoring' is often used in the context of education to describe a combination of coaching, counselling and assessment, where a classroom teacher in a school is delegated responsibility for assisting pre-service or newly qualified teachers in their professional development.

Mentoring means guiding and supporting trainees to ease them through difficult transitions; it is about smoothing the way, enabling, reassuring as well as directing, managing and instructing. It should unblock the ways to change by building self-confidence, self-esteem and a readiness to act as well as to engage in ongoing constructive interpersonal relationships. Mentoring is concerned with continuing personal as well as professional development (CPPD) and not just continuing professional development. In the process, personal and professional values come under scrutiny and are subject to change. Mentoring is not synonymous with cloning because it means developing individuals' strengths to maximize their professional and personal potential and also that of students who come under their care within a classroom situation. The principal stakeholders in mentoring are the mentor and mentee but others share responsibility. The mentee is as much an agent in bringing about effective mentoring as the mentor and neither can operate in a vacuum. Mentoring is about whole-school and whole-HEI (higher education institution)/school partnerships.

The present government initiative to introduce training schools with mentoring at their core represents a realization that good practice in teaching comes from good training in good schools.

## Managing change through mentoring

If we accept that mentoring is about easing transitions and ensuring development, it follows that it must be responsive to the individual strengths, values and needs of both the mentor and the mentee. Mentoring is a process whereby skills change. People will change too as they interact with new colleagues, as mentoring involves assisting with transitions in thinking as well as transitions in actions. When the labour market in which teachers are employed was more stable, before the era of job sharing, short-term contracts and teacher supply agencies, trainees entering our profession tended to remain in one teaching post or at least in one school for many years. Today the picture is becoming one of ever shifting employment with teachers moving between schools for promotion. Many government initiatives increase the feeling of flux, so mentoring has become a must in the process of educating for change. A mentoring relationship enables teachers to cope with change and enables change within teaching too. Mentors and mentees create new knowledge about teaching as they interact together and this new knowledge also leads to change.

## What mentoring was

The story of the archetypal mentor in Homer's *Odyssey* is recounted in most of the literature about school, business and employment mentoring. What commentators fail to point out, if they even appreciate it, is the social engineering, the determined manoeuvring and the changing relationship of Mentor (or more correctly the goddess Athena who impersonated Mentor) who undertook social, moral, spiritual and cultural mentoring with Telemachus. Athena decided to intervene in Telemachus's life – he wasn't given a choice about this – because she felt an obligation to support his father, Odysseus. Whether you draw on the 1881 translation from the Greek manuscript offered by Alexander Pope or the 1941 translation by Rieu, the story is quite clear about Mentor's (Athena's) role in mentoring. She has no intention of producing a dependent clone but she has very precise objectives in mind for the young Telemachus:

> I will go to Ithaca to instil a little more spirit into Odysseus's son and to embolden him to call his long-haired compatriots to an assembly and speak his mind to that mob of suitors who spend their time in the wholesale slaughter of his sheep and his fatted cattle.
> (Rieu)

> Meantime Telemachus the blooming heir
> Of sea-girt Ithaca, demands my care:
> 'Tis mine to form his green unpractis'd years
> In sage debates, surrounded with his peers,
> To save the state, and timely to restrain
> The bold intrusion of the suitor train;
> Who crowd his palace and with lawless power
> His herds and flocks in feastful rites devour.
> (Pope)

Mentor sets out to educate Telemachus, to support and inform him so he may rid the good-for-nothings squandering his father's estate, but Mentor has plans that go far beyond the reorganization of Odysseus's homeland:

> You – my friend – and what a tall and splendid fellow you have grown! – must be as brave as Orestes. Then future generations will sing your praises.
> (Rieu)

Mentor is not above manipulating – but strictly for Telemachus's good! She tells the boy that she thought his father was at home when she first meets him. She is prepared to impersonate him to gather a suitable crew and vessel to take him on his travels because he is not yet strong enough to do this for himself. She allocates the manageable task of collecting provisions. As Telemachus grows in physical and mental strength and determination, so she eases out of the role of instructor and trainer to be a more distant support and enabler, eventually leaving Telemachus to fight without her. As Telemachus grows in maturity, psychologically and physically, so the nature of Mentor's (or rather Athena's) mentoring is changed. Mentor gives Telemachus challenges of increasing complexity as he gains confidence and support is reduced.

The original Mentor may have even slipped in an occasional lie but in the context of school-based mentoring there must be honesty and openness if a trusting relationship is to emerge.

Mentoring, within the context of this handbook of good practice, is seen as a dynamic process whereby a teacher new to the profession not only learns the necessary skills (as an apprentice carpenter might for example) with a more experienced colleague but also develops the attitudes, practice and knowledge that are conducive to bringing about pupils' learning in class. Mentoring is about developing the 'person' of the mentee and it thus entails interaction, flexibility, and response and sensitivity on the part of the mentor. Ensuring that trainee teachers can demonstrate their ability to meet the burgeoning number of standards for qualified teacher status is only part of the picture and though it is essential, as we shall see in later chapters, it is only part of the process of supporting and preparing a new teacher's emergence as a professional educator. Similarly, mentoring impacts on the mentor in a clearly two-way process of learning alongside the mentee.

Mentoring is 'not done to' the mentee but, rather, it is 'done with' the mentee, because the mentee and the mentor are both actively engaged in the process of gaining understanding about teaching to improve it.

## How this book is organized

There are six principal parts to this handbook:

1. Developing your mentoring.
2. The process of mentoring.
3. Mentoring in a school context.
4. Contexts for mentoring in school.
5. Developing competence through mentoring.
6. Useful references for improving mentoring.

Each of the first five parts is divided into chapters to assist you in considering related aspects of mentoring. The sixth part is dedicated to useful references, to enable you to read more widely and learn more than this handbook can possibly tell you about what constitutes good practice.

There is a glossary towards the end of the book to assist you in identifying the meanings of the acronyms used in the text. Acronyms are in very wide use in higher education (HE) and school-based initial teacher training (ITT)!

The chapters within each of the first five parts have a common format within four main sections:

1. Overview of the chapter.
2. The main text of the chapter.
3. Illustrations of good practice.
4. Summary of good practice arising from the chapter.

Within each chapter there are illustrations of practice in mentoring. Some are examples of good practice and others are examples of how not to approach mentoring. They are stories taken from encounters with real mentors in real schools. Some are intended to be brief illustrations of a particular point within a chapter, others are more lengthy case studies. For the illustrations there is a commentary to assist you in seeing what might be indicative of good practice. In the second chapter of Part One, 'How you can develop the teaching skills you already have', the stories of a teacher's progress to mentoring are used in a different way. You, the reader, are invited to interact with the narrative, to create your own story about your own teaching history and to use this as a basis for identifying how you can bring your unique skills to mentoring.

One way of using the chapters in a formative way for your own professional development would be to create a file with a subsection relating to each of the parts of this handbook. Use the file to frame a portfolio of your development as a professional educator of trainee teachers. Include notes about each section, and keep a diary of your reflections on your own practice and use this as a basis for target setting, action planning and evaluating your mentoring.

Mentors often find that it is helpful to begin exploring their role by reflecting on one of the many models of mentoring that have been developed by researchers to represent what they do. These models are symbols, ideals perhaps, and through analysing them, accepting some parts and rejecting others, new understandings about the nature of mentoring can evolve. Having a model of mentoring in your mind can assist you in sharing a culture of mentoring with colleagues and to develop a shared language with other educators to express what you know.

Daloz's (1986) model for example sets out a two-dimensional representation of the interaction between support and challenge within a mentoring relationship. Fromm (1970) speaks poetically about the paradox of mentoring:

> [The mentor] does not give in order to receive; giving is in itself exquisite joy. But in giving he cannot help bringing something to life in the other person, and this which is brought to life reflects back to him; in truly giving, he cannot help receiving that which is given back to him . . . In the act of giving something is born, and both persons involved are grateful for the life that is born for both of them.
> (Fromm, 1970: 24–25)

For Anderson and Shannon (1988) mentoring can best be defined as

> a nurturing process in which a more skilled or more experienced person, serving as a role model teaches, sponsors, encourages, counsels, and befriends a less skilled or less experienced person for the purpose of promoting the latter's professional and/or personal development. Mentoring functions are carried out within the context of an ongoing, caring relationship between the mentor and the protégé.

For Hale (1999), mentoring unblocks ways to change by increasing self-confidence, self-belief and action orientation by building interpersonal relationships. Maynard and Furlong (1994) describe a triple model of mentoring: the apprenticeship model and learning to see; the competency model, which relies on systematic training; and the reflective model where the focus moves from teaching to learning. All models differ yet all usefully stimulate thought about mentoring.

## Mentoring within initial teacher training

When Kenneth Clarke, the Secretary of State for Education in 1992, announced that schools were to assume the role of teacher training that was previously organized, assessed and validated almost exclusively by lecturing staff in higher education institutions, he effectively created a new workforce – the school mentors. Prior to Government Circular 9/92, which heralded the shift to school-based training, and clause 12 of the 1994 Education Bill

> The governing body of any county, voluntary or maintained special school, or of any grant-maintained school, may provide courses of initial training for school teachers.

many classroom teachers had been called upon to supervise the day-to-day welfare of trainee teachers during their periods of teaching practice in schools – where they were expected to apply the theory they had learnt in the HEI to the practice of working with children. From the findings of five research studies carried out into mentoring in initial teacher education, McIntyre and Hagger (1994) identified a worrying trend that

> dangerously, there is evidence that many teachers, experienced in supervising student teachers on conventional school practices, anticipate little need for change or for new learning as they become mentors . . . such complacency is profoundly misguided.

For mentoring goes far beyond supervision – mentoring is about active education of the mentee. Given there is no existing word - 'educatee' perhaps

– to describe the educational nature of the programme for the neophyte teacher we will have to continue the use of the term 'trainee'. However, mentoring goes beyond training – beyond instructing and coaching – all of which suggest a predetermined goal with definable competences. Mentoring in a school-based context involves training but it should be more, much more than just this, for it is the means of education! It is about drawing out learning and development as well as feeding in instructions.

When would-be teachers apply to undertake the Post Graduate Certificate in Education, they have just 36 weeks to make the transition from being a full-time student or journalist or shop assistant or managing director of a firm, to being a member of a team of educators who are supposed to 'know' what to do in schools. Obviously, some come more equipped in terms of mindset and employment experience than others for making the transition. Any kind of career transition is likely to be dramatic and often challenging and sometimes can also be frightening. School-based mentoring should smooth and enable this transition.

It can no longer be assumed that the school-based mentor will be older as well as more experienced than the novice teacher. The increasing incidence of career changes, some voluntary and some enforced, mean that a young mentor who has been in teaching for a relatively short time may be mentoring a colleague 20 years older; someone who is moving from a position of expertise and experience in a different field of employment. Mentors have to be more experienced than their mentees but not necessarily older. Mentees do not come as a blank slate, for they bring their employment and learning history with them.

## Relating teacher training to mentoring

As the trainee moves through stages from novice to professional educator, the mentor's own practices and values will come under increasing scrutiny by the trainee. At the beginning of this introductory chapter we looked at what mentoring is and is not. At different stages in the trainee's development it is likely to be appropriate for mentoring to shift in emphasis and to concentrate upon giving information, coaching in particular skills, and modelling skills for the trainee to replicate. For the trainee to become a 'professional' the mentor has to move on, too, by constantly creating opportunities for drawing out the mentee's potential to teach. Only then can mentoring be considered to be truly educative and creative.

## Selecting suitable mentors

It seems self-evident that not every teacher can be an effective mentor. Just as it is impossible for all graduates to be effective schoolteachers (even if they wanted to be) so it takes a particular constellation of skills, attitudes, knowledge and action to enable even an expert classroom teacher to become an expert mentor. However, pockets of ignorance remain and there are those in even senior positions in schools who seem unaware that some teachers cannot and should not be selected to become mentors. It may be true that every teacher is potentially a mentor, but not all teachers can necessarily be really good mentors!

To be a good mentor there must be personal and professional engagement. Although mentoring is not synonymous with counselling, which requires specialist training, there are times when it seems similar. Teachers should bring to mentoring a willingness to listen and to support but not an overwhelming drive to solve others' problems. Anyone considering mentoring should carry out a self-audit, but where do you as a mentor start? The following questions might help:

- What do I bring to the mentoring relationship?
- How can I communicate what I know about teaching to my mentee?
- Am I prepared and able to be more than an instructor and a coach?
- Am I willing to open my own practice to the scrutiny of a trainee?
- Am I willing to support, challenge and educate beginning teachers?
- Do I understand that the dynamics and focus of my role must change?
- Am I prepared for the sacrifices, changes and challenges this will entail?

Looking at the list of desirable characteristics for would-be mentors to embody can be a very daunting prospect for a classroom teacher who is already well aware of what mentoring might entail but is willing to undertake this rewarding and demanding responsibility. It is even more daunting for classroom teachers who find – and this happens quite frequently – that they are the only ones to mentor in their schools. They have no choice.

Where a classroom teacher is asked to become a mentor there should be a shared responsibility from the outset. The senior management team should not approach the teacher unless whole-school support is available, which goes beyond just providing time for mentoring and possible remuneration. The teacher also has a responsibility to decline if mentoring would jeopardize the welfare of the pupils in their classes and their own health. Mentoring is a very demanding activity and never to be undertaken lightly. A mentee's career is at stake and so is the education of the pupils in his or her care.

## Good practice in the selection of mentors

Before any commitment is made to offer a school-based placement for a trainee, the headteacher has to decide if engaging in initial-teacher education is in the best interests of the school at this particular point in time. With an OFSTED inspection due or an application to become a language or technology centre in the offing, it is all too tempting to neglect the very particular needs of trainees or to hand them more responsibility than they can realistically handle. Mentors need to be prepared for their role. Can they be spared to attend mentor development sessions in school and, where available, at the higher education institution? Mentors need support; they need mentors too. It is unrealistic to expect a teacher to become an expert mentor in one short training session. There should be a network of support in school for the new mentor – a sharing of ideas, skills and personal support that puts professional development of all staff at the heart of what a school does. Where one mentor cannot realistically undertake, for whatever reason, the range of mentoring activities that any novice teacher needs, the mentoring should be shared within a subject department, or across a school in the case of primary placements. In this way novices can be given the chance to receive the breadth and depth of support that they need. Some mentors may be skilled in using information and communications technology (ICT) and others less so. If they work together with the mentee there is a potential for all to gain professionally and personally.

How long should teachers teach before they become mentors? Long enough to be sufficiently experienced to know the needs of a new teacher and know how to meet these. Long enough to have gained sufficient seniority in the school to understand how to facilitate the new teacher's development there and to have the power to enable this. It really isn't a question of the number of years because some mentors, like some classroom teachers, can be experts in a matter of a few years – while for others it takes longer and for some it is unlikely to happen at all. For mentors to be facilitators they must have the strong support of the senior management team. Mentors should *never* be selected purely upon the grounds of how this might shake them out of a stupor of professional sloth! When novice teachers are asked to identify the single most important skill for a mentor to possess, they tend to choose the ability to listen. In the survival stage it is tempting to offer quick-fix tips but this alone does not move the trainee to become a professional. The mentor must be capable of listening, sizing up the situation and offering appropriate action, advice and, sometimes, silence. Novices need space to make mistakes, to reflect, to grow and to improve. The period of their education in schools requires space to experiment yet also requires guidance so that pupils are not put at risk.

The process of selection of mentors must crucially empower them to be effective by ensuring a supportive context in which they can undertake their work. There has to be willingness by the senior management team (SMT) to acknowledge the contribution of the mentor's work to the school and to offer constructive criticism too. The mentor has to be capable of working in a team and alone, while contributing to the professional development of someone whose values may be different.

## Training mentors

The processes of educating pupils, and trainee teachers and mentors, have much common ground for both are learning situations. Basic mentor training may cover the 'mechanics' of mentoring; how to set up a teaching timetable for the mentee; how to observe a lesson in relation to criteria set out by the training programme; how the assessment of a novice is decided upon. But mentors need more than this. The realization that they need to challenge as well as support is all very well in theory and with a strong student it is relatively easy to bring about. Problems arise when the student is not (yet) strong or needs to develop certain skills. There is a well-worn phrase within teaching that says that good classes will virtually teach themselves. Neither good classes nor good trainee teachers can do this and so it is crucial to train mentors to bring out the best in them.

With a move away from a situation in which colleagues in higher education had almost sole responsibility for novices' assessment, there has been a move towards making this role a central tenet of mentors' responsibilities. Higher education tutors cannot expect, or be expected, to 'sort everything thing out' when they visit a trainee who is encountering difficulties in a school. They are there with a responsibility to support the trainee and the mentor – but above all they are increasingly becoming moderators. They can point out what needs to be done in terms of targets to be set and actions to be undertaken but they do not have the training responsibility. Since mentors have become assessors they may seek the guidance of higher education colleagues but they must take action to resolve problems. Obviously, then, mentor training needs to encompass strategies for dealing with problems, and to recognize and resolve potential problems before they erupt.

Training needs to develop a mentor's interpersonal skills and in particular to enable sound and appropriate communication with adults as learners. The fundamental principles for doing this are related to those employed in good classroom teaching but they are also distinct. Good mentors will realize that they need to diminish the power differential between mentor and mentee and, although novice teachers may err like wayward pupils, telling one off like a child is not very productive! There has to be resilience, drawing upon compassion for a colleague's mistakes coupled with an acknowledgement that mistakes are inevitable. The art of mentoring lies in recognizing which mistakes are acceptable, which are not and where a trainee with problems is failing and, with that, the skill of knowing how to advise a failed trainee to leave. On the other hand, a mentor needs to learn how and when to praise. Mentor development is about engendering a sensitive balance between giving personal and professional development and learning to assess a fellow professional consistently and fairly against a predetermined set of criteria. But good practice mentoring is more than this, as the following account of mentoring shows in a letter written by a trainee teacher:

> You allowed me to be me and that was all I was looking for. I know you, being you, will tell me that I am not giving myself enough credit in this process, but essentially what I am saying to you is this, I have always known what was inside me, but no one has ever taken the time or had the skills to draw those qualities out before. The insight you have given (facilitated) me into myself has literally turned my life around. I know that of course I am not yet quite 'a professional' and that I have still got a lot of work to do in terms of subject knowledge, planning, class-management, differentiation, but the key thing is, I now feel I have the ability to tackle all these things and more.

## Summary

Mentoring is:

- active education of the mentee;
- concerned with continuing personal as well as professional development;
- about easing transitions and ensuring development;
- responsive to individual strengths, values and needs of mentor and mentee;
- compassion, and acknowledgement that mistakes are inevitable;
- recognizing which mistakes are acceptable and which are not.

# Part 1. DEVELOPING YOUR MENTORING

Chapter 1

# Structured mentoring

In this chapter we shall be considering the following aspects of mentoring:

- The need for structure in mentoring.
- Structuring the mentoring session.
- Stages in mentoring.
- Phases in mentoring.
- Identifying good practice within a structured approach.

## The need for structure

Although mentors are almost universally agreed about the intrinsic value of mentoring, the greatest difficulty cited is finding enough time for the mentoring process. Although a mentor may be allocated at least one hour's protected time, free from teaching or from being called in to cover for absent colleagues, there is rarely enough time to fulfil what mentoring requires or to fulfil it as comprehensively and conscientiously as good mentors wish. Why is this and what can be done about it? It comes down to prioritizing and choosing what to focus on. Sometimes the protected mentoring time is not made available and the mentor is then expected to carry out a role that deserves more status and support than snatched minutes here and there in a busy week. It could be that the senior management team (SMT) does not have adequate staffing to allow for time to be made available. If this is so, one must question very seriously whether there should be a whole-school undertaking for mentoring at all. It may be that the members of the SMT have not been involved personally in mentoring and are unaware of its demands. Possibly the mentor is not sufficiently skilled at managing time efficiently so what could be covered in a shorter time becomes a protracted process that grows as demands mount. Where there is a trainee who is encountering problems settling into teaching, a single hour is certainly inadequate because mentoring begins to grow until it exceeds the time that *is* available.

Frequently, the problem is a combination of several factors outlined above – and certainly there are some times in the school year when mentoring is going to be far more time consuming than at other times. So the managing of mentoring becomes paramount; managing by the SMT, but fundamentally by mentors themselves. Mentoring is always a compromise and so it is vital to negotiate the best compromise possible. School-based mentors must give of their best – simultaneously – to both their teaching in class and to their mentoring.

## Enabling a proactive approach

Good mentors actively manage their mentoring, by planning, evaluating and orchestrating their mentoring sessions. Good mentoring does not just happen. It is the result of prioritizing, sound management and structuring. This is not to say that each and every aspect of mentoring should be preplanned. Mentoring must operate flexibly to enable the mentor to be reactive where necessary within a proactive framework. But how is this balance to be achieved?

Once it has been agreed that there is appropriate staffing to undertake good quality mentoring, or enough staff who have the potential to develop into good mentors, the pressing priority is to consider the impact on the whole school. Mentoring changes the dynamics of the classroom and also the staffroom because it has a direct effect on pupils' learning in the classroom and on the ethos of the school. Effective mentoring is a whole-school issue. Consideration of the impact on the school must be the first priority – but the teachers and pupils *are* the school – so the degree to which mentoring has an impact is directly related to all of the personnel involved.

Once the decision to engage the school in mentoring has been made the critical eye must look to the timetable. Mentoring takes *time* as well as expertise – to be effective it needs quality staffing and quality *time*! Think of it as a priority alongside classroom teaching.

Mentors do tend to give their all to 'make it work out', but when they give their own time to help a new colleague they can burn themselves out and endanger their health and their classroom teaching. This is not structured mentoring; at best this is just coping. It is a deficit model where the mentor will be stretched by even a strong mentee and quickly exhausted by one who encounters substantial problems. Mentoring must be managed in order to protect mentors, trainees and their pupils. Sometimes, then, the mentor's provision for a new entrant to the teaching profession is endangered. Trainees often make significant sacrifices to become teachers and they always deserve the very best training. In this next section we are working on the basis of a protected hour's mentoring session. The numerous moments of mentoring that occur in passing, during break and between lessons, are essential but additional to this.

## Structuring the mentoring session

Good mentoring occurs where there is participation by both mentor and mentee and it takes place across three time zones: past, present and future. There is a parallel process for learning to teach and learning to mentor. You should begin with planning, thoroughly and meticulously, and as you become more confident and skilled use the plan as a springboard for your work. One of the reasons that mentors are so short of time is that SMTs often do not allow time for this planning. One hour *cannot* be sufficient unless some of it is allocated to the mentor's own planning and reflection in preparation for the coming mentoring sessions. Mentors effectively need nearer two hours recognized for mentoring each week: one hour for the mentoring and one for planning and reflection. The most proactive phase of mentoring is likely to be during the early days. A case can be made for managing the mentoring over the year, with more time allocated at the outset and less towards the end of the training. Mentor and mentee will know one another better and the emphasis can shift structuring the mentoring sessions so that they can fulfil the professional and personal needs of both parties.

## Stages in trainee teachers' development

Furlong and Maynard (1995) set out stages in the mentees' development within initial teacher education. Again they are explicit that the mentoring relationship does and must change as the mentee becomes self-sufficient. They look at how 'typical' novice teachers pass through five broad stages of development, reporting on research among trainee teachers. The stages are not necessarily identical for all trainees and depend on individual personality traits and professional experience but, broadly speaking, there are five stages through which all trainee teachers pass. The stages are only intended as a guide. Some novice teachers are at different stages when they start training, they do not necessarily develop in a linear way that the term 'stages' suggests. Some novices seem to regress in a second school placement; context matters.

My own experience as a mentor and as a tutor supporting mentor colleagues leads me to believe in a slightly different model from the one that I will set out below. For me, the division between the levels is never clear cut. I find that a trainee can be at the early idealism stage in one aspect of their work while simultaneously at the moving on stage in another. So much depends on prior experience in the workplace and the ability of the trainee to manage change and to develop.

The flexibility of good mentoring lies in recognizing where a trainee is in relation to a particular skill. It is vital for the mentor to keep open the relationship in a trusting secure way so that the trainee feels valued; just as we do with pupils. Learning to teach can be a very painful process as well as an exciting one, and we need to start our mentoring at the point where trainees are, rather than where we think they should be at a certain point in the training programme. If they so far 'out of sync' with the programme that they do not have a chance of successfully completing their training it is important to explain this to them with compassion and to value them as people.

Because it is easier to structure mentoring if you have a concept in your mind, I have set out below the model that Furlong and Maynard have created for us. It is a very useful starting point but do bear in mind my point about linear development. Some trainees will be experienced teachers already, whereas others might never have been in a classroom in the UK before undertaking their training. All have their constellation of personal values and this constellation will determine, to a large degree, how effectively the mentoring process is.

Not only is it important to give a structure in mentoring – it is vital to ensure that it reflects the values your trainee brings to the mentoring relationship and the values that you, too, hold as a teacher.

### *Stage one: early idealism*

According to Furlong and Maynard this stage had the following characteristics:

> Teaching and learning were commonly seen in a simplistic way. Students maintained that it was their relationships with pupils that would be the crucial factor in terms of their effectiveness as teachers . . . [Their] role was seen as something that just happened without a great deal of effort on their part.
> (Furlong and Maynard, 1995: 76)

### *Stage two: personal survival*

> . . . students at this stage tended to be reactive rather than proactive – leaving the children to define the situation.

While students stated that they wanted 'their own personality to come out', many maintained that if they didn't emulate the teacher's style, the children would not respond to them.

The 'ideal' teachers they initially wanted to be had been replaced by the teachers they had to be if they were to survive.

### *Stage three: dealing with difficulties*

> . . . students try to replicate or 'mimic' what they believed to be teacher behaviour. This attempt to gain at least a 'procedural' understanding of what it meant to be a teacher left students feeling vulnerable.
> . . . for many students their worth as a teacher and as a person appeared to be judged by how far the pupils appeared to 'like' them.
> . . . even though hours were spent on planning students seemed reluctant to differentiate the work they devised in terms of pupils' abilities.

### *Stage four: hitting a plateau*

> . . . their teaching still showed little appreciation of the relationship between teaching and how children learn.
> . . . students started to gain confidence in their abilities to manage classes . . . However, most students were still 'acting' like a teacher, rather than 'thinking' like a teacher.

### *Stage five: moving on*

> Student teachers needed to be 'moved on' to understand the role and responsibilities of being a professional educator.
>
> It is only when students are forced to face and to reconsider their own value position, what that means in terms of their practice and in particular in terms of children's learning, that there is likely to be any 'movement'.
>
> . . . the importance of helping student teachers to evaluate their beliefs about the nature of teaching and learning is fundamental if students are to develop into fully professional teachers.

Central to Furlong and Maynard's study is an indication about how an understanding about the stages of trainee teachers' development can inform mentoring that is sympathetic and conducive to progress. It is likely that trainee teachers pass through stages similar, if not identical, to those identified by Furlong and Maynard, so it follows that mentoring must also pass through stages too. It must evolve not only as a professional relationship to teach and refine skills but as a personal relationship to support the professional aspect that is essential to 'learning teaching'. It would be inappropriate for a mentor to challenge trainees in the first stage of development to 'think like a teacher' when they cannot yet understand what 'being a teacher' entails. Similarly, trainees who are 'hitting the plateau' cannot hone their skills if left without challenge. Being 'good enough' to stand in front of a class and deliver an aspect of the scheme of work is not sufficient, for if they are to understand how pupils learn, trainees will need to learn to plan towards maximizing their pupils' learning potential in class.

## Phases in mentoring

Let us consider teacher training in three phases. The relative length of the phases may well vary, not only according to the number of weeks spent in school but also according to the profile of the mentoring programme to be tailored to a particular mentee's needs. The length of the phases of the mentoring suggested below may not coincide with the length of school placement, so phase one may be longer for trainees encountering problems than for those who have fewer difficulties.

### *Structured mentoring: phase one*

This is the time for the first of several mentoring audits and this needs to be carried out well before the mentee arrives. Once the staffing, possibilities for shared mentoring and the allocation of resources to mentoring are agreed within the school, it is time to find out as much as possible about the trainee who will be joining your teaching staff. The Graduate Teacher Training Registry (GTTR) forms are used by interviewers as a basis for deciding whether or not to invite a candidate for interview. These contain a vast amount of useful detail as a basis for tailoring a programme. The forms are vitally important but they are only one part of the jigsaw of information a good mentor needs to tailor an appropriate mentoring programme.

In selecting an applicant for interview, there should be two major considerations. These are knowledge (subject and teaching/pedagogical content knowledge) and personality. Pedagogical content knowledge (Shulman, 1986) means the know-how to turn subject knowledge into a form that can be taught so pupils can learn. The range of knowledge possessed by the trainee should be apparent from the GTTR form. Look for the content of the degree, check that the candidate has the equivalent of a 'C' grade at GCSE in maths and English (and also science for primary and middle school applicants). Finally, check the class of degree. Most training institutions look for a 'good' honours degree, that is a first or second class. If the degree class is not given you should contact the candidate and ask.

Look for evidence in the letter of application of teaching experience or working with children – how else do these candidates know that they want to teach? Are these well-informed applicants or will they need an induction, especially if they are coming from another country? Will they understand the pastoral responsibilities in schools and have they personal experience of pastoral care in schools? Look for any evidence of disability – dyslexia, perhaps, or visual impairment. How will the school build on the candidates' strengths in order to empower them? How will the school provide a remedial or booster programme if candidates need this?

Next, consider the GTTR form for any indicators of the candidate's personality. The official reference should provide clues, but it does not always do so. Not all referees take as much professional consideration as they should in providing a reference. Is the reference from a colleague who really knows the applicant and how far does the reference coincide with the content of the application? If this is a GTTR form that appears to be good in every aspect except the reference, consider seeking information from the second referee. Once you know about the trainee from the documentation there are two other sources of information to be used as a basis for constructing a tailored programme before the trainee arrives in school. One is observation; consider inviting the trainee to join your department prior to the start of the training course for a day so that you can get to know one another. A second is discussion with the interviewer. Did the interviewer keep notes during the interviewing process and did they write a summary on a pro forma that you can look at to form a clearer impression before your trainee's arrival? Of course, first impressions are important, but they can be misleading. Be open-minded when you meet trainees for the first time and make allowances for them to overcome initial nerves and feel at ease with you. Trainees will come with preconceptions about teaching and as these are likely to colour how they perceive schools and teaching you should plan to work from what is positive in those preconceptions while preparing yourself to challenge what is negative.

Starting from the hour of protected time that is made available for mentoring, determine your objectives and those of your trainee. In the early part of the training you are likely to have a more didactic role where you should determine the content of the sessions while allowing adequate time for the trainee to bring their own concerns and questions to discuss. Think about the hour in terms of three sections and for each section allow opportunities for input from you and from your trainee teacher. Before the session, make provision for recording the process and outcome. You may decide to make carbon-copy notes as you proceed and give a copy to your mentee after the session to keep on file. Mentoring can fall down when it becomes overly prescriptive or overly reactive to a succession of the trainee's crises.

The three main sections of each mentoring session – and you may decide initially to allocate some of your designated time to each – are feedback, present considerations, and planning.

### Feedback

Begin by putting the trainee at ease, explain your objectives for the session and ask the trainee to summarize his or her own. There will be crises and time must be allocated to deal with them but they should be addressed in relation to the three components of the hour: feedback, present considerations, and planning. Plan the order in which you both will cover your objectives within the three phases. It may be helpful to ask trainees to bring a list of three objectives with them, drawing on outcomes and targets of a previous session if possible.

### Present considerations

The second part of the mentoring session should look at issues that the mentor knows to be part of the school-based and, where appropriate, higher education-based training programme for the mentee. In this part it is essential for the mentor to enable the trainee to bring together the salient issues covered in higher education-led sessions and integrate them into what is happening in school. It is not a case of theory there and practice here – rather it is a matter of integrating theory and practice in both sides of a partnership programme. In the case of school-centred initial teacher training (SCITT), there may be HE involvement but the same applies – the issues covered in the middle part of the mentoring session should serve to integrate what is covered in other parts of the training (in general professional studies as it is often called) within a subject context. For example, an early theme in the training curriculum for mentoring might be classroom management, so the mentor needs to be familiar with what has been covered in all aspects of the trainee's programme to ensure that targets set in the subject area are grounded in non-subject coverage.

### Planning

The third part of a mentoring session should be devoted to planning on the basis of what has occurred in the previous two parts. The objective for this part is to plan precise steps that will enable trainees to appreciate their own progress and that will provide a basis for mentors to monitor, assess and record mentees' progress in their teaching. Where the standards for qualified teacher status (QTS) are likely to be covered in parts one and two of the mentoring session, the third will involve planning opportunities for the trainee to consolidate achievement in the standards. This is where the craft of mentoring should be most evident. This third part of the mentoring session should be appropriately challenging to the mentee's level of development and conducted within a secure haven.

### *Structured mentoring – phase two*

The components of the mentoring will be similar to phase one – feedback, present considerations, and planning –but this second phase is characterized by a shift in emphasis in the proactive role of the mentor and trainee. Whereas in phase one it was the mentor who led the mentoring, planning proactively and reacting as needs arose, in phase two a balance between trainee and mentor is beginning to show. The mentor will still need to plan what is to be the theme of the sessions – but the trainee's task is to bring issues for discussion about this theme, drawing on lessons taught and on targeted observation of others' lessons.

This phase may well coincide with the block of teaching in a second or complementary school placement. However, many trainees do regress a little when they move from one teaching environment to the next and so phase one may well overlap into the start of the new school. It is vital to ensure good communication at the handover between the mentors in the two school placements. It is just as important, however, to bear in mind how different every school is and that trainees may well demonstrate different characteristics in their teaching in a new context. Phase two mentoring is geared to widening a trainee's experience as well as deepening it. There are likely to be relatively fewer completely new experiences than there were in phase one. In phase two, the trainee should be consolidating what he or she has already begun to learn. The mentor is therefore charged with the responsibility of creating opportunities for this consolidation. During this phase, the trainee will probably apply for teaching posts. Offering a mock interview with appropriate feedback is particularly useful as is identifying areas where the trainee has not yet made much progress. This phase often coincides with plateauing, so it is important to support and challenge the trainee who now needs to learn the interaction between teaching and learning, between teacher and pupil, by planning a mixture of stir and settle activities.

In this second phase the trainees will need to consolidate and extend their level of understanding about assessing pupils' work. They may find that it is confusing to have to learn a new way of recording marks and so it is important to explain why things are done in a particular way. It is crucial to ask why the trainee does things in a certain way. This phase is the time where trainees need room to reflect on new teaching strategies and meet examination classes if they have not already done so. They should not be given full responsibility for pupils' progress but it might be useful to use the trainee's support for preparing for exams.

By the end of phase two the trainee would be expected to have covered all aspects of the competence statements relating to qualified teacher status. This should have included opportunities to gain experience in using ICT as a teaching tool to enhance and stimulate pupils' learning during lesson time. It is essential to set trainees the task of using ICT within a minimum of at least three lessons where they can demonstrate that, without this medium, pupils could not have learnt what they did. It may be possible to begin the use of ICT in the first phase but, certainly, it must be under way in phase two of the course if the trainee is to be able to demonstrate competence by the end of initial teacher training (ITT).

During mentoring sessions the balance of support and challenge will shift as the trainee is encouraged to question the mentor's teaching. This can be a daunting experience for a novice mentor and it is important to realize that this questioning is an important stage for the trainee.

---

## Illustration

W is a strong novice teacher – so promising that his main school placement has appointed him to a teaching post for the coming year. He is very hard working and conscientious – if anything, he works too hard. In lessons he tends to keep to 'stirring' activities so that neither he nor the class has much respite from active teaching and learning. His mentor is becoming rather concerned and his visiting tutor agrees that W needs to learn to pace himself or full-time teaching will soon exhaust him.

## Commentary

Although W is well able to manage a highly flamboyant and very active teaching style at present, his mentor realizes that this is not a suitable way to allow him to develop without a check. If W continues at his present rate, not only is he likely to become tired but his pupils will be deprived of an opportunity for reflection and managing their own learning too. He needs to develop variety in his repertoire of teaching strategies and with appropriate advice from his mentor and from his HE tutor he decides that an adjustment is necessary. The professional tutor is supportive and invites W to watch how he paces himself for monitoring assessment and recording pupil progress in class

By the time W is ready to leave this second school placement he and his teaching style have matured considerably, without his losing enthusiasm. In fact, he has even more energy at his disposal as less time is taken for marking after the school day is over because some now occurs in lessons. During mentoring sessions W is keen to learn how to pace himself better and so he brings questions to the sessions as well as waiting to respond.

---

### *Structured mentoring: phase three*

This third phase is in some ways the most exciting of all. The trainee is beginning to demonstrate independence as a teacher and, although still in need of support and challenge, is moving into a more autonomous phase of development, ready to enter our profession. In the mentoring there is a sense of polishing, of completing a task after the early stages of training and of working towards full-time employment. This is sometimes called the 'extension' phase and it should be the time when mentors can ease back on the proactive aspects of their mentoring. It is a time for redressing the balance of the early mentor-intensive phases of ITT. The trainee is not yet qualified but he or she can be expected to take increasing (though monitored) responsibility for teaching. By now the trainee is to be regarded as a member of the teaching team in school – no longer quite a novice but not yet quite a teacher. Now is the time to assist trainees in refining their style and scope of teaching, to ensure that they have met consistently and confidently all the requirements of the externally imposed standards of teaching competence. Where there are gaps in competence these must be met as rapidly as possible and it is likely that often the mentoring sessions will be targeted by the novice teacher.

Within the mentoring it is important to retain the triadic structure we began with: feedback (how has the trainee been getting on recently?), present considerations (issues for the mentor and trainee to raise), planning (as a result of discussing the two previous sections here).

---

## Illustration

In the third term of the PGCE year C the trainee teacher has completed a period of teaching in a second school and has returned to her home school with a series of targets to meet prior to gaining QTS. Her mentor, B, has agreed to video record some of C's teaching so they can concentrate on improving her skills in those specific areas set out by her second school mentor. B is not necessarily present during the recording – in fact she prefers not to be because she would like to enable C to build her working relationships undisturbed by her presence in the classroom. C watches the video of each lesson and then gives it to B to watch and prepare from in advance of mentoring sessions. C decides which standards for QTS she wants B to look at in relation to the videotaping.

## Commentary

B is in a strong position to enable C to avoid just plateauing. C is a competent teacher and unless B were able to monitor her progress she might not move on to becoming better. By videorecording the sessions B and C can discuss C's progress in specific areas with a range of classes and over time. They do not record every single lesson but at most make a weekly recording. C does not always find remembering what happens in a lesson easy and in her first term tended to get very emotional if all did not go according to plan. Now she can see where she is learning to adjust her teaching and where she needs to be even more responsive to avoid a likelihood of problems. C is particularly happy in this school placement because she knows that she is receiving individual attention, even though B has many calls on her time. At the end of the training year B passes these videos on to the university tutor for her to use for mentor development in future years. The videos provide an authentic resource for discussion and B is on hand to talk to other mentors about how her mentoring changed over the year.

---

Questions to ask yourself:

- Why is structuring a mentoring session important?
- How are you structuring your sessions at present?
- How might you improve the structure over the period of training?

Good practice within a structured approach is:

- drawing from your trainee's values and experience as starting point;
- the result of sound management and planning by the mentor;
- phased and structured across the training programme;
- organized in a cycle of feedback/present considerations/planning;
- integrating theory and practice.

Chapter 2

# How you can develop the teaching skills you already have

In this chapter we shall be considering the following aspects of mentoring:

- How good teaching relates to mentoring.
- Identifying mentoring skills within a teacher's narrative.
- Auditing your teaching skills as a basis for mentoring.
- Identifying good practice in developing teaching skills for mentoring.

What is good teaching? What are the skills that good teachers bring to their work? Is it even possible to define categorically what teaching is? As Nicholls (1999: 6) says: 'the complex nature of teaching and education could not be unravelled by a simple categorisation of qualities, knowledge and skills.'

This handbook is not intended as a 'handbook on good teaching' – it is designed to be used as a springboard for creating and disseminating good practice in mentoring. However, you cannot separate school teaching and school-based mentoring without overlooking the richness that their inter-connection can bring. The multiple focus within this chapter sets out to help you to develop an awareness of the teaching know-how you already have as a basis for mentoring.

In this first section, you are invited to interact with the text by reflecting on the narrative of your own teaching while you read an account of someone else's. As you read the illustration, look within your own practice as you look at this teacher's for those skills, values, aptitudes and knowledge that you can constructively apply to your mentoring. What you should seek to do is to create and then investigate a 'virtual text' about yourself, using your own knowledge and understanding about yourself as a teacher. By doing so you can begin to view your teaching and your potential for mentoring with slightly more detachment than just 'thinking' gives you.

## A teacher's story: part one

> I am teaching in a middle school where there are about 400 pupils aged between 9 and 13. I have class of 29 children between 10 and 11 years old. They come from a mixed catchment area – some affluent, middle class, from the housing area nearest the school, some from the docklands brought up each day by bus to our former secondary school. There are four year-groups and I liaise closely with the head of my year as a form tutor and run my own department across the school.
>
> This is a middle school and so, although I am the head of French, I am teaching several subjects including needlework, English, humanities, art and environmental studies. I have an honours degree in French and Spanish and a PGCE that has begun to equip me to teach modern languages and environmental science in a middle-school context. I run a mini-enterprise club for pupils at lunchtimes, growing plants for school bazaars and selling lettuces to the school! The French teaching presents no challenges in terms of my subject knowledge; the real challenge is how to motivate the children in my class. For some, holidays in France are a regular excursion in summer, but for others France might as well be Timbuktu – it's foreign and irrelevant! I realize early in my career in this school that if I am to make any headway I have to make studying French appealing and I have to pitch my lessons just right or the pupils just glaze over like earthenware pots – and 'blob'. I have to motivate pupils across a range of abilities.

What teaching skills does this narrative highlight?

- A depth of subject knowledge (honours degree).
- Addressing individual needs (motivating mixed-ability groups).
- Pastoral as well as curricular teaching experience.

What personal skills support the teaching?

- Commitment (extracurricular activity).
- Enterprise (running a small business with the pupils).
- Versatility (teaching a number of very different subjects).

What else? (What personal values, what knowledge about how to teach my subject etc?)

Now create a narrative of your own teaching and ask yourself what teaching skills your narrative highlights and what personal skills support your teaching.

## A teacher's story: part two

It's almost 10 years on now. I have worked in a second middle school as head of French and as head of year. I am now employed in a huge secondary school as a main-scale teacher of French. My career has not followed the path I expected, from head of French on finishing my probation to becoming a member of a senior management team. I am registered disabled; I have difficulty walking and driving long distances because of a back injury. I have been ill for seven years and I have missed about a year of full employment. It's my teaching that has kept me going. It is still my passion and there are days when being with the kids pulled me through.

I have a tutor group of 30 lively 12-to-13-year-olds. I take some of them for French – they are put in sets according to their ability in maths, largely, and some are taken by other staff across the year group. I am taking GCSE French for the first time. I am also taking a special needs group for French every Monday morning and last thing on Friday. They are a challenge to my teaching – getting the level just right – and to my patience – every step we make together seems to be untrodden each weekend! I have a bottom set French group in year 11 – what a group! Though they are studying supposedly in parallel with my own set in year 11 we are talking survival skills with one – and potential university entrants with the other. Although I am extending my teaching skills I am narrowing my teaching scope to one subject, French. I am learning how to work on an individual level with the SEN group, creating individual education plans. My own disablement means I cannot rely on physical presence to lead my classes – I am no longer physically dynamic. Time to develop more dynamism in my voice and approach . . .

What teaching skills does this narrative highlight?

- Subject knowledge and knowing how to teach it (teaching to GCSE level).
- Planning teaching and classroom management.
- Monitoring and assessing learning (individual education plans).

What personal skills support the teaching?

- Determination (to continue with a career despite disablement).
- Self-awareness (realizing how impatience needs curbing).
- Positive outlook (time to develop . . .).

What else?

Now create a further narrative of your own teaching and ask yourself what teaching skills your narrative highlights and what personal skills support your teaching.

## A teacher's story: part three

> Three years on and I am working in an upper school for pupils aged 13 to 19. Appointed two years ago as a main-scale teacher to teach French up to A level I found myself assuming the role of head of Spanish when the then postholder took early retirement. To revise and upgrade my language skills I've negotiated one morning a week during my first year here at a local university, to join the advanced Institute of Linguists Course for Spanish. After a year I start studying for an MA in 'Teaching Modern Languages through Language and Literature' and I am studying how to teach as I do it – teaching literature at A level. Taking on the head of faculty's role I am not given a form tutor group at first but I join a night-school class for German – we have three foreign languages here and as head of faculty I feel duty bound to know all three. I begin to master using a computer – slowly and impatiently as I prepare my MA dissertation for marking. I train to become a mentor in the Licensed Teacher Scheme in our county and my first mentee arrives. She is a French national and teaches up to A level in her first language. Her German needs upgrading but she is learning as she teaches it to GCSE.

What teaching skills does this narrative highlight?

- Subject knowledge and knowing how to teach it (teaching to A level, ICT).
- Taking responsibility for subject development (Institute of Linguists and MA study).
- Developing management skills (head of faculty).

What personal skills support the teaching?

- Negotiation (I negotiated one morning . . .).
- Inquisitiveness (I am studying how to teach . . .).
- Commitment (I train to become a mentor . . .).

What else?

## Summary

In the brief sections of narrative above we have looked at how one teacher has progressed through her teaching career. As we have considered each part of her narrative we have reflected on different aspects of her teaching skills. How might these provide the basis for mentoring?

This teacher has brought many skills to mentoring both personal and professional;

- A passion for teaching and an ability to enthuse.
- An ability to structure learning.
- An awareness of the need to take responsibility for professional development.
- An understanding that all learners are individual and learn in different ways.
- A willingness to adapt her subject knowledge to teaching at different levels.
- A realization when she needed assistance in working on subject knowledge.
- A profound knowing that while being a teacher she is simultaneously a learner.

Which teaching skills needed development for her mentoring?

- Working with adult learners.
- Recognizing stages of mentoring in tune with the stages in novice teacher development.
- Developing observation skills as a basis for mentoring feedback sessions.
- Making planning more explicit as a basis for explaining to someone else.
- Identifying a variety of teaching strategies to adapt and from which to choose.

We have looked at a teacher's narrative to try to appreciate how teaching skills can underpin mentoring. In the previous section of this chapter we carried out a retrospective auditing process of identifying teaching skills transferred to mentoring. In the next section we will begin a forward-looking audit. Rather than asking what skills might be useful in mentoring, now identify skills you already have that are potentially transferable but that need development.

In the final section of this chapter a framework is provided for you to audit skills you might bring from your teaching into your mentoring. As you consider each category, think not only of your aptitude for mentoring but of the practical steps you will take to improve and develop.

There is a simple scoring mechanism for each category below, which works like this. For each category award yourself a score representing the amount of a skill you think you possess:

5. very high;
4. high;
3. enough but no more;
2. lacking;
1. non-existent.

Subject knowledge. How far do you have the knowledge to answer pupils' questions securely and to teach at least to A level in your subject?

Pedagogical subject knowledge. How far can you adapt what you know so that pupils can understand your teaching easily?

Autonomous skills. How far can you work effectively without the need for other teachers to motivate you and ensure your work is accurate?

Intuitive skills. How far are you accurate when you just 'read a situation' and how far can you adjust positively?

Creative skills. How far can you bring your imagination to bear on even the most mundane tasks so they become enjoyable and a basis for more creativity? (I remember one teacher who organized his class into an army regiment – and the registration was never quite the same for me after that!)

Team cohesion skills. How far are you an effective member of a year and subject team? How much do you understand about the senior management team's (SMT's) role in school?

From this audit, profile yourself as a potential or improving mentor:

5s. Time to celebrate! That is what you are aiming at as basis to become an expert mentor.
4s. Look to improve on your 4s as you engage in mentoring – keep going!
3s. Maybe undertake team mentoring so your mentee has expert input other than yours. Any opportunity to get yourself a mentor to help improve your skills?
2s. Are you ready to undertake mentoring yet? Probably not.
1s. Maybe you should not even be teaching!

If you are scoring 3s in all areas think about team mentoring so you have time to learn in tandem with an experienced mentor. If your subject knowledge needs attention take steps to improve it but consider how you can contribute to mentoring in the meantime – by giving personal support, by form-tutor mentoring and by mentoring across generic skill areas such as classroom management, lesson planning and assessment in class.

## Positive steps for your own development

- Audit your teaching skills by using the standards for QTS.
- Ask a friend or colleague whose opinion you respect to audit your teaching.
- Remember that honing your professional and personal skills can take time and effort.

### *Why should you trouble to audit your skills?*

If you are to be not only an effective teacher but an effective mentor this needs insight into where your strengths lie and where there are areas that you must attend to for development. The potential of mentoring is becoming widely known, not just as a part of initial teacher training in schools but as a mechanism for effecting change and improvement at whole-school level.

As the DfEE Consultation Document (0008/2000) states, there is an increasing realization at government level, as well as in schools, of the potential of mentoring

> as a key professional development activity. Current practice varies and teachers have found benefits from mentor relationships . . . Teaching assistants sometimes have a mentor teacher within their school . . . in many cases the relationship offers the mentor as much opportunity to develop and learn as the person being mentored. We should like schools to develop this excellent practice.

There is an increasing emphasis within the teaching profession on enabling teachers to meet professional standards of practice. This is the basis of the move towards performance-related pay advocated by the present government. In order to meet these professional standards, teachers are being encouraged to produce teaching portfolios to demonstrate their achievements within each standard. Mentoring offers teachers the opportunity to include such evidence in a way that enhances their professional standing. In the final section of this chapter we shall look at two further illustrations in the light of this interest in extending mentoring within a school to enhance the mentoring of trainee teachers. In both cases the mentors could well include evidence of their mentoring in portfolios as an extension of their teaching expertise and their professional development.

As you read the two illustrations below reflect on the teaching and personal skills that are evident and consider how they have been developed as a basis for good practice. Use the same techniques we practised in the first part of this chapter when we looked at the teacher's story. Consider the teaching skills that each narrative highlights and the personal skills that support the teaching.

## Illustration

P has been a mentor for some years now. She has a very high level of subject knowledge in her first subject, French, and studies hard in her own time, as well as by taking groups of pupils to study centres in Germany to improve her second language teaching skills. The pupils respect her and like her – she has a wonderful sense of humour, which motivates them and keeps them on task in class. Above all, P has developed sufficient self-confidence in the presence of a class to allow them to become largely autonomous. She encourages them (after considerable modelling and with plentiful guidance) to manage the teaching of one module of their course-work syllabus to their own class. She sets them high standards. On a personal front she realizes that it is important that she continue to learn how to improve her teaching. The senior mistress is acting as her mentor and gives her active guidance in how to develop her practice and they have both agreed that she should learn to teach a third subject in school. P is sharing her own learning experiences as she engages in mentoring a trainee teacher.

Consider the following:

- P is keen to develop her teaching skills.
- She is sharing with her mentee the difficulties she is encountering in learning.
- She inspires because of her humility – 'I don't know everything and never will.'
- She has a certainty that, suitably prepared, everyone can be increasingly autonomous.
- She models her belief in the mentoring process by sharing her own experiences.
- She ensures that the novice teacher will feel that he is not isolated in learning to teach.

## Illustration

B is a teacher and an experienced mentor. Her enthusiasm is contagious when you meet her. She immediately puts you at ease by her manner – she has well-developed skills for working in a close-knit community and, in addition to acclaim from all her pupils, has been graded 1 by OFSTED. She works by guided discovery – creating experiments in her science lessons so that the pupils can draw their own conclusions and advising so they make informed observations. She is a very polished practitioner who is a sufficiently accomplished teacher to invite her trainee into her classes and give a running commentary on why she is teaching in a particular way. Because her subject knowledge is so strong and her teaching skills are so advanced she can devote less attention to preparing for the pupils' lesson and more to preparing for guiding her novice through the lesson as a learning experience. She clearly sees teaching the pupils as her first priority in class but uses the 'space' where she encourages pupils to become autonomous to advise and mentor her trainee quietly in a corner. The pupils accept that everyone has to learn.

---

Consider the following:

- B has learnt to integrate her teaching and her mentoring so that everyone benefits.
- By openly commenting on her teaching she picks out those things she is doing well.
- B highlights where she needs to pay attention so the pupils will learn more effectively.
- She is mentoring each pupil to develop their potential as learners. She extends.

In this chapter we have looked at how you can and should identify teaching skills in preparation for developing mentoring skills. It is important to realize that learning to mentor is as much a lifelong experience as is learning to teach. It entails an integration of both personal and professional skills and it is likely that you will need to develop new skills in order to mentor.

Good practice in developing teaching skills for mentoring means:

- Reflecting on the skills evident within your own history of teaching.
- Auditing your own professional and personal skills as a basis for development.
- Considering the skills you are going to need because you are meeting new challenges.
- Looking for ways to adapt, develop and apply your teaching in mentoring.

Chapter 3

# Mentoring for your own professional development

In this chapter we shall be considering the following aspects of mentoring:

- Developing mentoring to become a better mentor.
- Developing mentoring for professional and career advancement.
- Developing mentoring to enhance interpersonal skills.
- Identifying good practice for integrating mentoring and professional development.

> A considerable amount of research into mentoring has focused on descriptions and conceptions of what mentors say they do, and what they actually do. There is much less on what it is possible for mentors to do.
> (McIntyre *et al*, 1993)

The most exciting aspect of mentoring is how it changes not only the outlook of your mentee, but your outlook too. One benefit is that you will be able to see perspectives on familiar situations that you could not previously see, because you see through another's eyes. This enables you to reflect on your own personal and professional ways in a way that is normally impossible. Through reflecting on your own practice as a teacher, you can learn, with the help of your mentee, to observe what you do. Because of this you can begin to see where improvements are needed. Mentoring fills in the reflexive blind spots we all have. By showing you another viewpoint you can see yourself and your teaching in a fresh light and this is invigorating and challenging, while providing you with insights into how you can improve.

Effective mentoring is a proactive process that necessitates change and needs appropriate support if it is to be successful. In chapter 12 we consider how mentoring is a whole-school issue and, to adapt Eric Hawkins' metaphor from a modern languages context, to say that mentoring without sufficient professional support from colleagues is akin to 'gardening in a gale'. With appropriate support in school, mentoring can be a springboard to personal and professional development. Depending on your motives and context in undertaking mentoring, it may provide the basis for promotion to a post in senior management and/or the start of a course of study towards a higher degree. Much depends on your motivation as the mentor and the application of your developing skills, knowledge and beliefs about teaching. Seeking promotion or advanced study may not be your goal. Mentoring in itself will bring you a sense of self worth that enhances your wellbeing in a profession where praise for staff is rare.

## Becoming a better mentor

As Brooks and Sikes (1997) rightly point out, it can be problematic to define a good mentor let alone a 'better' one. What constitutes 'good mentoring' in one context, with one mentee, does not necessarily transfer across and so one comes to a vague definition of mentoring as 'the most appropriate means of engendering a trainee's effectiveness as a teacher in a given context'. Good mentoring results in good teaching by trainees, which brings good learning by pupils. However, a trainee unsuited to teaching will never be a good teacher and a 'good' mentor will give sound counsel to this effect – that the trainee should discontinue training – difficult as it is.

Having been a mentor, I feel that there are some hard and fast 'rules'. Surely the first 'rule' is to be a listener, the second to be attuned to a novice's needs as well as to the context in which mentoring occurs. 'Better' mentors are always looking to improve their listening skills and the nature and quality of the support and challenge on offer. The goal to aim for is improvement by asking yourself questions like Whitehead's (1989) 'How do I improve my practice as a professional teacher-educator?' and seeking to develop those areas you know to be wanting.

Although there are some references in the burgeoning literature on mentoring, a limited amount has been written about how expertise in mentoring can be used as a basis for improving other aspects of teaching. Among these few inspirational sources of professional guidance available to mentors seeking to be more effective is an article in McIntyre and Hagger (1996) by Bush, Coleman, Wall and West-Burnham. This offers useful guidance for mentors seeking to extend and apply their skills to other aspects of school life including peer mentoring, where a pair of mentors undertake mentoring for and with one another; mentoring as part of an appraisal process in a formal or informal context; mentoring with a view to possible promotion – to become a middle manager or a senior member of staff; and educating pupils to mentor one another to improve academic achievement, perhaps, or behaviour. Peer mentoring for example can enhance one's understanding of the difficulties as well as the rewards of management – difficulties sometimes hidden or overlooked by those not directly involved in management.

Who will train the next generation of teachers? We should be considering how teacher training will be different from and better than the present way of doing it. It stands to reason that good teachers should be trained by good mentors in good schools. Mentors are in a strong position to train teachers and to train the next generation of mentors. They can complement the role of higher education colleagues that have a breadth of overview that comes from working with mentors across a number and variety of schools. If research into mentoring is their area of interest they are likely to have a depth of knowledge that most mentors would find hard to match. In the initial training of teachers and ongoing professional development, experienced mentors and colleagues in higher education institutions have much to contribute. One challenge facing mentors occurs when they are no longer called upon to undertake mentoring. The saddest outcome would be to allow skills acquired to lapse and there really is no need to do so. Ghaye and Ghaye (1998) invite teachers through systematic reflection to make sense of their teaching context, values and practice and to consider how to contribute to school improvement – what better starting point than from their mentoring?

How can the professionalism of teaching be enhanced? How can practising teachers contribute to its development through the General Teaching Council? Mentors are among the best placed to respond here. If teaching reform is to be appropriate to good practice it should surely be good practitioners who are researching it and shaping it, as well as traditional policy makers. 'We currently need to find new narratives for redefining education so that schools can restore a sense of purpose, tolerance and respect for learning' (Watkins, 1993).

We need narratives of successful teachers talking about their work so that skills are understood and become the basis of future policies for shaping teaching and learning within our schools.

## Mentoring for professional and career advancement

When novice teachers start teaching they experience a change not only in skills but also in their knowledge and beliefs (Calderhead and Elliott, 1993). Similarly, mentors develop skills beyond those they have honed in the classroom and their knowledge and belief system can be changed too. As trainees show new ways of approaching teaching, mentors review their own practice and as a result, subtle shifts in both perspective and practice are likely to occur. Teaching in the classroom may become largely implicit, almost automatic, and it can take the 'Why?' of a mentee to spur a teacher to reassess the reasons behind a particular action and see a choice to be made – to continue if it is appropriate or to shift ground or abandon if no longer so.

Sometimes it is only when there is an extra pair of hands in the classroom that the teacher (now mentor) has moments to reflect on his or her actions while in the classroom. Reflection after teaching may well occur but significant moments can be lost to memory. Sometimes reflection results from seeing a novice who is having difficulties – perhaps with class management – and this challenges a mentor to question his or her own practice through seeking to explain how and why they continue to do what they do. Through questioning the mentee's practice, mentors will often confront their own teaching and through educating a neophyte they will often remember that they are also learners. Ironically, as a mentor is mirroring the mentee's practice so the mentee reflects back the mentor's. This is an opportunity for constructive professional development resulting from this new insight. Although it can be daunting to be challenged about your practice, and uncomfortable to be observed and questioned closely about what you are doing, it is offering you an opportunity to a review and enables you to develop as a professional educator.

## Becoming a better teacher

How can I use what I learn in mentoring to improve my own practice? Having the opportunity to work with a beginning teacher is a privilege and can give insights into how adults and pupils learn and the nature of their preferred learning styles. Armed with these insights, the mentor is in a strong position to offer more flexibility in classroom teaching and, as the opportunity arises, for colleagues. Knowing that learners learn in very different ways reminds us all as teachers to focus on flexibility and creativity in planning a learning experience for others.

As mentors, we can change how we talk to ourselves in our inner dialogue. By listening attentively to the messages novice teachers feed themselves, as learners we can examine our own internal messages and take steps to make the messages positive and constructive. As we are party to our novice teacher's intentions, hopes and goals in teaching we have an opportunity to lay our own alongside – are they still relevant, are we reaching our own potential and if not quite, what steps can we take? As you action plan with a novice, action plan for you!

Excellent teachers may opt for promotion and run the risk of spending more and more time away from the very teaching that they excel at. Fortunately, some decide to remain in the classroom to give of their best. Mentoring can refresh and enliven their approach as they share the insights of their experience. Mentors can and do decide that they have much to offer by diversifying their teaching as a result of mentoring. Seeing the novice explore and experiment with new teaching strategies can inspire and refocus expert teachers. As a result some decide to produce better teaching resources to use – there are exciting opportunities for publication. Others develop their school role and successfully take over responsibility for staff development.

Contact with university staff is often cited by mentors as one of the benefits of their work but this contact has potential that is often not exploited to the full. Higher education colleagues can bring different but complementary skills to the classroom from those of practising teachers. For a start they have the opportunity to work in research on a regular basis. It should be that collaborative research can bring mutual benefit to both parties – perhaps addressing problems of underachievement in class. Moreover, lecturers have access to literature and library facilities and are well placed to be advisors beyond the immediacy of a single school. Feiman-Nemser in her 1998 paper on teachers as teacher educators concluded that the most effective mentors

> did not learn to mentor on their own. The most effective British mentors in the study met regularly with university colleagues to talk about what novices need to learn, how they can be helped to learn that, what role the university and school-base mentors should play in the process.

Few would deny there can be a difference of culture and of language between schools and HEIs but it is through collaboration that there can also be a growth in understanding and a bridging through communication. As school-based mentors and HE colleagues work together they can come to develop a shared understanding and from this research projects of direct relevance to raising pupils' achievement in the classroom can take shape. Sharing in undertaking research means that this will be of relevance to the classroom and can begin to change the culture of waste and superfluity where potentially useful research is shunned by or unknown to teachers.

## Becoming a university tutor

University tutors are likely to have been practising teachers in schools. As teachers become lecturers it is often the research component of their new post that is the most daunting prospect. Most teachers who make the transition are pleased to have done so, but stress levels are rising in higher education institutions and the traditional view of academics with ample time to think and carry out research is changing as the pressure is mounting from the implementation of research assessment exercise cycles every five years. It can take two years to see something you have written appear in print and matching writing to the style and audience of a journal can be tricky.

A university lecturer needs the ability to contribute in depth across a range of commitments.

Teaching in a university setting, within a four-year B.Ed or one-year PGCE course, is sometimes available where, for example, a tutor is successful in 'buying time out' to undertake research. A transition from classroom to university teacher may entail a part-time teaching commitment in both locations or on a temporary secondment. Working in a university setting can open up possibilities that are unavailable in school. True, there is the loss of contact with pupils and for some tutors this is a painful loss, but it can be balanced against exciting opportunities for teaching adults and the chance to work at the cutting edge of research; and the flexibility of working schedules is an attractive incentive to many teachers in schools bound by the clock.

Moving from teaching in school to teaching at university can open up new ways of using your mentoring skills. When I arrived to start lecturing I was assigned a mentor who had been teaching in schools and working with student teachers for several years. What I valued in particular was her ability to see how to enable me to develop my thinking and to create openings for me to further my professional development. We grew alongside each other and you can see how useful this experience was and the stages we went through together in Fletcher (1997a).

In the illustration that follows we can see how opportunities for professional development arising from mentoring are not necessarily planned but can arise spontaneously in practice.

---

### Illustration

This situation occurred fortuitously when a university tutor was taken ill just before some of the teaching block visits were to take place. There was no substitute tutor available and rather than delay the visits to a small group of novice teachers in one subject area the university invited a group of experienced mentors in the PGCE partnership to peer mentor by visiting trainees in each other's schools. Each mentor visited two others and undertook the role of the subject tutor, with which they were all familiar as they had already been working together for some years.

## Commentary

This was an excellent piece of in-service training for everyone involved. Where there was any question about a novice teacher's progress a later visit was booked for the university tutor to act as a moderator. The novices were delighted to be monitored by a 'real teacher' who had immediate credibility as a subject expert and professional expert. Visiting mentors were accorded status not only by the schools they visited but also by colleagues in their 'home' school. The mentor whose novice was under review welcomed the support and challenge of a colleague. In fact the subject tutor gained too – one visiting mentor said she had no idea that it was such a demanding job! She thought that the mentors in school would automatically accept a visitor's assessment – and she learnt that they don't and that they (very rightly) demand justifications.

Having experience of mentoring from the two perspectives of the school and the HEI would be indicative of positive professional development for an employer considering job applications.

---

## Becoming an LEA advisor

Mentoring skills in a subject area are often similar to those used in an advisory context. When advisors work closely with a teacher in school they may well engage in an action research enquiry, identifying a problem, looking for possible strategies to lead to better teaching, selecting a course of action, undertaking it and then assessing the outcome. This, in short, characterizes many activities in active mentoring. Mentors take on a co-enquirer role as they work with novice teachers to develop skills. There are times of course when the mentor instructs rather than enquires – in tasks where there is one correct or accepted effective practice. There is little opportunity for co-enquirer stance in learning how to complete a class register – some aspects of initial education are best handled in a straightforward didactic way. Others, which involve designing learning tasks for pupils, for example, are better met through co-enquiry and benefit all participants in terms of openness and creativity. The LEA advisor is looking to develop teachers' own teaching skills, and this goes beyond simply 'telling them how'.

Short-term release on contract is sometimes available for practising teachers to develop their career within advisory work. There are often times when advisors will call on experts in a particular teaching technique or who have developed a new approach to a challenge. Working with pupils who have particularly acute special educational needs is just such a case. These teachers are a valuable asset to the advisor for in-service work in schools because they model existing good practice and talk creatively from experience about ways of overcoming problems. Advisors are well placed to support mentors' induction tutors' work in ITE.

## Becoming an OFSTED inspector

For some teachers the very name 'OFSTED' brings goosebumps to the skin and a lump to the throat! There are good inspectors and less good – just as there are good and less good teachers. A mentor might undertake training to be an inspector without the commitment to inspecting schools other than his or her own. Training is challenging and the skills that equip a qualified inspector can be put to good use within a school. Often schools employ a local education advisor or independent consultant to undertake a pre-inspection audit of strengths and weaknesses as preparation for the full inspection. A mentor who knows how to work with adults to improve their practice and has an in-depth knowledge of the OFSTED assessment process will already be a valuable resource to his or her department and to the school as a whole.

## Gaining accreditation in mentoring

As yet there is no national accreditation for mentoring but it may be that the present professional framework for teachers will some day encompass this aspect of many teachers' work. The resistance to such accreditation is understandable – teachers do not want to sacrifice their flexibility in teaching by finding that they face more competence statements as mentors. For some schools, becoming a mentor is still a bolt-on extra within job descriptions. Should it become an established and evaluated part of a teacher's role it will have a social, political and economic dimension. Undoubtedly, mentors deserve overt recognition for their work and national accreditation would lead to a recognized currency on the job market. Given this recognition, it should be easier for mentors to argue for time to carry out their work – and perhaps for financial reward, although time is a commodity that most mentors feel to be more important. Accreditation would be a way of ensuring quality assurance and raising mentoring's status through professional recognition at national level. Too often mentoring relies on goodwill!

As an extension to mentoring, it is important to compile a portfolio of evidence that shows how you have influenced the development of your trainee's practice in a constructive way. Portfolio assessment is at the basis of the present government's move to performance-related pay and it is likely to be only a matter of time before this is introduced for mentors too. In due course there will surely be standards for professional development for mentors, so be prepared!

In the next illustration we can see how this teacher has already aligned her thinking to refection by taking part in reviewing and rewriting the course programme for initial teacher training. Because she is used to working with HEI colleagues, she already knows much of the 'language' of acronyms and she has a shared culture through collaborative work. She has also networked with lecturers who will be keen to help her develop her writing at MA level.

---

### Illustration

J is an excellent teacher and has already established herself as a likely candidate for promotion in the near future. However, she would dearly like to gain a further academic qualification soon to bolster her chances of selection. She has taken a major role in co-writing the subject programme of the PGCE course with the university tutor and she has assisted on several occasions with interviewing prospective trainee teachers. When asked why she does not register to study for at least one module of an MA programme she will say 'I don't know what to say – I don't have the language to write essays – what would I write about?'

### Commentary

J has taken positive steps to meeting her desired goal. By already working closely with the university she has a very sound insight into both sides of the partnership's operation in initial teacher education. She has also voiced her concerns to the colleagues who can assist her. Rather than shying away from what she perceives as a struggle she has sought advice. What J is seeing is that she can and should research her own practice as a mentor and increase the understanding of other mentors and would-be mentors as well as HEI lecturers in education.

Berrill (1994) rightly pointed to the lack of mentors writing about their work in schools. If existing literature about mentoring is lacking, mentors could contribute while they are still engaged in mentoring (Fletcher and Calvert, 1994). Alternatively they are well informed and well placed to share their knowledge about teaching. Colleagues in higher education are only now coming to the realization that 'emotions are at the heart of teaching and that good teaching is charged with emotion' (Hargreaves, 1998). In a competence-bound context good teachers affirm that, although competences are useful, teaching is more than meeting competences.

## Mentoring for enhancing interpersonal skills

In the course of their mentoring, some mentors realize that their particular contribution lies in their ability to offer personal and professional counselling to their mentees. There are courses in counselling available at night school and others are organized by professional organizations looking for volunteers. Mentoring is about sharing, supporting and advising and the insights gained can be of immeasurable use to others. Mentoring activity can be closely allied to counselling. Should mentors decide to extend and consolidate their counselling skills by additional training, this in itself is likely to extend their range of mentoring capabilities. In a paper entitled 'The Mentoring Dilemma (1998, Mentoring and Tutoring) Brian Gay and Joan Stephenson set out four possible models of the mentoring relationship – mentor: protégé; master: disciple; craftsman: apprentice; and therapist: client. It seems to me that the 'ideal' mentor should be flexible and knowledgeable enough to cross these transitions, and one goal for CPD in mentoring could be extending one's skills to encompass another 'approach'. The client: therapist model is non-judgmental, open contact in terms of time and supportive yet questioning in its application.

## Developing professional relationships through mentoring

Mentoring relationships have individual differences, so in developing mastery across a wide range of interpersonal skills a mentor will be better able to assist mentees from dependence to the establishment of their own knowledgeable and skilful independence. The design of a mentoring programme is often seen in terms of surface arrangements and contracts, of content knowledge, where it is in fact the underlying relationships that determine its success. As the relationship changes over time and with experience, it can be difficult to maintain a sound relationship. This challenge is perhaps at its greatest when the mentor in school, the friend and supporter, is called upon simultaneously to become an assessor. Mentoring in schools is often characterized by this dilemma, where the mentor is supporter and judge, and it takes patience and skill to work well.

Mentoring demands good interpersonal skills for working with adults so that even the most difficult messages can be communicated to a mentee who naturally feels vulnerable, self-doubting and at times even jealous of the apparent ease with which you manage a class and timetable. Relationships are at the heart of all classroom teaching so a good teacher will be actively seeking to enhance these as a process of lifelong learning. Beyond the classroom contacts between teacher and pupil, as a mentor you can only benefit from learning how to interact in an assertive but non-combative way in the microcosm that is the staffroom.

Questions to ask yourself:

How well do I work with others?
How can I further improve my interpersonal skills?

Good practice for integrating mentoring and professional development:

- Mentoring does change perspectives and offer scope for growth.
- Mentoring can enhance mentoring skills and understanding.
- Mentoring can enable professional development within teaching.
- Mentoring can enable professional development outside teaching.

Chapter 4

# Researching your own practice in mentoring

Why should I undertake research into my own mentoring? What have I got to say? Would anyone else be interested in me? Have I got time? Can I write well enough? I would do but . . .

In this chapter we shall be considering the following aspects of mentoring:

- Good reasons for undertaking your own research into mentoring.
- Ways and means of undertaking research into mentoring.
- Undertaking research about mentoring in collaboration with an HEI.
- Good practice for undertaking research to enhance your mentoring.

## Why undertake research?

It is a simple but daunting fact that most of the research about mentoring that has been published has been written by people who have never done it . . . Some of it is excellent nevertheless, but some of it does not represent mentoring as a mentor knows it is. Why? Perhaps because mentors tend not to tell anyone else – or at least anyone beyond their school – what it is like to mentor. In fact, most mentors do not even talk about their mentoring to anyone else on a day-to-day basis. They do not share their successes and they do not share their problems. Should this concern you as a mentor? The simple answer is 'Yes!'. Why does it matter? If mentoring is to have status, to have the recognition it deserves, it must be understood. At present there is a dearth of understanding by policy makers. Mentoring, although widespread in teaching, has nowhere like the status it is accorded in industry or business. When was the last time you went to a conference for mentoring at a regional or at national level? Have you ever been to one?

Research is a word that seems to discourage teachers so let us examine what it really means. Research is finding out, finding out how it works, why it works, what makes it work – better! Poor research is obscure and tends to be written in language that is so ridden with acronyms and jargon that it is difficult to understand. The whole point of research into mentoring is that it can help mentors and potential mentors do their work better. But there is another vital aspect too. Imagine you are driving a car. There are blind spots for even the most competent driver. We use mirrors to enable us to see what is happening behind and beside us as we drive. Research can show you what you cannot see, what happens beyond the school you work in, and the potential that mentoring might hold for whole-school improvement. It fills in reflexive blind spots, as Brown (1998) calls them, by offering the full picture you need to see but cannot quite – yet!

Let us consider each of the questions posed at the start of this chapter:

- Why undertake research into my own mentoring?
- What have I got to say?
- Would anyone else be interested in me?
- Have I got time?
- Can I write well enough?

But instead of leaving them unanswered, this time let us supply an answer for each one.

Q. Why undertake research into my own mentoring?
*A. Because you know things that others do not about what it is to be a mentor.*

Q. What have I got to say?
*A. You might explain how you are helping someone become a teacher . . .*

Q. Would anyone else be interested in me?
*A. Yes – because you are unique and yet you share a common experience with others.*

Q. Have I got time?
*A. Have you got time to think about improving what you do?*

Q. Can I write well enough?
*A. If you can write reports for parents you can write a report on what you do. There are many ways of keeping an account of what you do that do not involve writing. You could*

*record a mentoring session on audio cassette or on video – review it and then jot down notes under 'Good points' and 'Points to consider'You could learn to write.*

Research need not be difficult to carry out but it is exacting and it must be rigorous and valid.

## How can I start?

You might decide to research for no purpose other than your own understanding However, if you do this there is a strong possibility that you will not be getting a 'rounded' view upon which to base target setting and appropriate action plans. Martin Buber (1947) wrote about the need to see perspectives other than our own when we are seeking to know about ourselves. He spoke of the need for a 'You' in addition to the 'I' being researched. Let's go back to the metaphor of driving a car and needing mirrors to cover blind spots (the feedback from another person enables you to see more about yourself). As a researcher you need to put what you find 'into the public domain' in order to validate it. Your findings need to be tested for validity through comparison with others' experiences.

It is ironic that we spend years learning how to make our teaching implicit – so we do not have to spend time consciously thinking about what to do next in the classroom – what Schon calls 'knowing-in-action'. Then as mentors we have to unpack what we do and why – for sharing. Before we can share what we do we need to be able to know what we are doing – to make the implicit things we do explicit – and research can help us understand what we do so that we can share.

## Undertaking research about mentoring in collaboration with a HEI

There has been a recent welcome upsurge in opportunities for research in collaboration with colleagues working in higher education. At one time it was considered to be an occupation only for lecturers or full-time research staff but gradually the agenda has changed and teacher-researchers are becoming more numerous. There is funding available at national level to support small-scale research and at international level for more ambitious collaborative ventures.

You are not alone if you feel apprehensive about carrying out research but if you allow this apprehension to deter you then you are denying yourself a worthwhile opportunity to develop. In the illustration that follows, the teacher was very unsure about undertaking research when she joined an MA summer school at the university where she worked in PGCE partnership. It wasn't until some days into the course that she discussed the fact that she had already been carrying out research into her own practice and was delighted to discover that this provided a very sound basis from which to undertake a study into her own work in mentoring at her school.

---

## Illustration

P has been a mentor for several years within the PGCE at a university renowned for the quality of its research. This reputation for research tended to conjure up visions of previous examinations and she made no mention of her own investigations throughout her mentoring years in conjunction with this HEI. In the past year, her school has been undergoing a period of staffing disruption and OFSTED had identified a number of weaknesses in its recent report.

P was singled out for praise for her teaching in this report and her work in supporting new staff was highlighted. Largely because of this aptitude she is an excellent mentor too, and decided to take a Masters module in the theory and practice of mentoring. Unbeknown to the tutor with whom she had worked for a number of years, she had been video recording her teaching in class as a basis for identifying how she can improve her teaching. She already has skills as a researcher but lacks some confidence in her ability to write. Eventually she confides in the tutor who reassures her that it is important that she passes on what she knows as a mentor. Happily, she takes up the offer of a place in the summer school week for an MA module in mentoring.

This is a wonderful story with a happy ending. P's confidence grows as she meets other mentors on the course and realizes what she has to offer. There are other mentors from partnership schools too, and although, at the outset of the week, they are resistant to the idea of writing an assignment, by the end they are supporting one another and fired up to start work.

P sends in an outline for the assignment after a few weeks back in school. She draws her study from the mentoring she has already done and the mentoring she is just beginning: a mixture of mentoring trainees and newly qualified teachers. She has decided to write a comparative study of the two forms of mentoring. The tutor reads her outline and offers advice and P drafts a fuller account. This is reviewed by the tutor who offers more advice and eventually P submits her final assignment for marking. It withstands rigorous assessment.

## Commentary

Like many teachers, P already has the skills and the background knowledge to enable her to undertake research into mentoring. The OFSTED inspection acts as a catalyst to her. She has to review her own work while acknowledging that she already has demonstrated considerable skills in her teaching and supporting other staff. Undergoing numerous staff changes in her department has meant that her mentoring skills are in ever-greater demand. She is already undertaking an enquiry into her work but however much she reflects on her own mentoring her audience will be limited to herself unless she shares the valuable insights that she has gained.

By joining the MA group she is able to learn from mentors from international schools as far away as the Bahamas, Cyprus, Bahrain and Kenya, and from other contexts: business and police work. She is able to gain a wider and deeper perspective on mentoring by discussion and by reading. As well as sharing insights into mentoring in her school she can contribute to building up a picture of mentoring in the United Kingdom along with other members of the summer school group, which includes several mentors also working locally.

Several higher education providers now offer MA modules in mentoring and others offer action research modules that form an excellent basis for exploring mentors' own practice in schools, with a view to improving it. It is possible to approach research in a number of ways: through telling a story and analysing it; through writing an historical account of mentoring and looking critically at how it has developed in the British education system. Perhaps the most relevant approach is to engage in researching through an action research in well-defined steps:

1. Think carefully about your work as a mentor engaged in school-based teacher training.
2. Identify the areas of your mentoring that you feel would benefit from improvement.
3. Decide which is the area of your mentoring you most want to improve. Write it down. Be as precise as possible about what you want to achieve. 'Must do better' is vague!
4. Brainstorm a number of ways that you might use to improve your mentoring and then consider if you have already tackled a similar challenge and succeeded. What have you learnt and what can you draw from your experience to help you meet this challenge?
5. Identify how you will evaluate your work if you undertake to improve it. How will you know that you are a more effective mentor and how will you prove it?

7. Choose one way of improving your practice that you can implement and evaluate.
8. Try it out.
9. Evaluate it.
10. If it works, move on to the next area for improvement in your mentoring.
11. If it doesn't work, go back and try out another strategy to improve what you do.
12. Take responsibility for improving your mentoring by seizing opportunities to improve. Talk to your mentee and to colleagues so you have a wider viewpoint.

---

## How do I do research?

Jean McNiff (1994) sets this out very clearly; I thoroughly recommend that you read the chapter entitled 'Making sense of the data' in her book, *Action Research: Principles and Practice*. I have reproduced a short extract below to help you get a feel for how to proceed:

> Having mapped out the general area of concern, try now to focus more specifically on something you feel you can do something about. Talk it over with at least one other critical friend. Jot down ideas. Writing ideas down is often far more cost-productive than gazing out of the window; writing sharpens the wit and focuses the attention. And there is something to be seen for your efforts, which is pleasing in itself.
>
> Keep the issues within manageable proportions; keep them relevant to yourself and people in your care. Do not be concerned at this stage with other people's problems. Sort out your own first.
>
> *Collecting the data*
>
> Data collection techniques fall into three broad categories:
>
> - Paper and pen methods (field notes/diaries/questionnaires)
> - Live methods (sociometric methods, interviews and discussions)
> - Ostensive methods (slide/tape presentations/DVD and video)
>
> *Which data?*
>
> . . . it is important to keep things in perspective and to remember that plans may well change as other issues are unearthed . . . Care should be

taken to store all the data until the time of the formal study is over. Certain issues may be latent and come to the fore only with the passage of time.

*Monitoring the data*

Keeping track of events is sometimes difficult, but essential if the study is to be systematic and legitimate when it comes to public attention. The methods of monitoring would be much the same as those recorded in the previous section on 'Which data?' but the personnel may well vary in monitoring. This may be done by the teacher-researcher, by a colleague or group of colleagues, or by the students themselves.

*Analysing the data*

Analysis is to do with making sense of what is going on in real life.

Making sense means deciding on what could be termed 'sense' in the first place, explaining why this rather than other actions are termed 'sense', and suggesting how the educative action in question approximates to the sense.

*Synthesising the data*

Synthesising the data means putting it all together in such a form that it may be easily communicated to, and comprehended by, other people. Analysis of an action research study implies identifying and agreeing criteria in action that can be used to explain what has happened or to indicate that improvement has taken place. Synthesis is how to explain the action in order to maintain it.

## Gaining accreditation for your research

If you do decide to undertake a programme of improvement for your mentoring why not seek an extra qualification while you do it? Increasingly, mentors are working with colleagues in schools, in conjunction with HEI providers, to acquire accreditation at an advanced study level. You do not necessarily have to write a long assignment. You can put together a portfolio of items that represent your mentoring experiences – and write a short commentary to explain what you have included and how it relates to what is already known about mentoring. You can see the process whereby portfolios are assessed against criteria at www.actionresearch.net.

This is what a professional portfolio might contain in addition to a detailed commentary explaining the relevance of each item to the study that is being undertaken by you, the mentor:

- evidence of your own prior professional development relevant to mentoring;
- evidence from a mentee's work that demonstrates your influence as a mentor;
- contributions you have made about mentoring to departmental or school policy;
- autobiographical accounts that explain your professional values and commitments;
- your own action plans for improving and extending your mentoring;
- evidence in a description and explanation of your professional learning as a mentor;
- a critical study of your professional reading relating to literature about mentoring;
- the application of research evidence to enhance your own or others' mentoring;
- videotapes and other visual and audio records that display values, skills and understanding relevant to good practice in mentoring;
- new action plans that show a modification of concerns, action plans and actions in the light of the evaluations;
- evaluations from colleagues that show how you have understood your own professional learning as you influence, improve and understand your own and your mentee's practice.

You may find that you can offer your research as a portfolio with a brief commentary or as a longer written assignment. In either case you will need to formulate a suitable research question that will form the focus of your study. This may come naturally to mind as you are engaged in your mentoring. On the other hand it may not, in which case it is wise to seek advice from an experienced researcher. He or she will be able to look with you at the context in which you work and help you to identify what it is about mentoring that really interests you, and will enable you to speak with authority after carrying out an enquiry. Often you know the general area in which you want to carry out research but identifying a precise question is rather more difficult.

## Possible areas of research to undertake into mentoring

- Are there similarities and differences in mentoring trainee and newly qualified teachers?

- What are the needs of new staff in international schools and how can mentoring help?
- How can mentoring staff and pupils bring about whole-school improvement?
- What kind of support do mentors need in schools to undertake their work effectively?
- What are the costs and benefits of mentoring trainee teachers at your school?
- How should mentors be trained and who should undertake this responsibility?

Advice about undertaking research should be readily available when you undertake advanced study at certificate, diploma and MA levels. Before you sign up for accreditation make sure you ask about how much individual attention you are likely to be given in framing your research. There are a number of HEIs offering accreditation in mentoring and the courses on offer may be run during twilight sessions, at weekends or in the evening, or during a vacation study period.

In the next illustration we shall look at a typical course of advanced study offered in late July.

---

## Illustration

This is the programme for the MA module in mentoring at Bath University. The course runs in the first week of the summer holidays and attracts a wide range of applicants from schools. By the end of the year following the sessions participants have to submit a 4,000 word assignment about mentoring or a portfolio with a 1,500 word commentary to explain what this includes. At present the Summer School involves a 20-hour commitment for university-based study:

This is how the programme is organized over the week:

Monday

What is mentoring and what are your expectations of this week's study in mentoring?

- models of mentoring;
- mentoring coaching and assessment;
- how to write an assignment.

Tuesday

What is so special about a mentoring relationship and how does yours operate?

- mentoring a pre-service teacher;
- selecting mentors in schools;
- matching mentors to mentees.

Wednesday

- good practice in mentoring;
- mentoring and assessment;
- mentoring resources on the Internet;
- planning a presentation (tutorials) on your mentoring and planning research.

Thursday

- analysing a videotaped mentoring session;
- mentoring a newly qualified teacher;
- mentoring in a business context – what can teachers learn from this?

Friday

- the prerequisites for effective mentoring;
- researching your own practice – and presentations from the group;
- looking at the future of mentoring – guest speaker from the TTA.

A busy and exciting week, which includes time for one-to-one tutorial discussion with your tutor as well as an induction into carrying out library-based research, not to mention a varied daily programme of social events if you are not too exhausted by all that hard work to take part!

---

## If you decide to stay in school and study

You may not need to leave your school even for a week to explore the potential of mentoring. Increasingly, in-school research is being funded by the TTA and advanced course modules tailored to schools' specific needs are offered in conjunction with local universities. Focus groups can be established so that issues relating to mentoring can be discussed among colleagues within the school – and with mentors from other schools. The possibilities of video-conferencing are just beginning to be realized in some schools – imagine holding mentoring conferences by this medium – getting advice and support on-line. As yet the National Grid for Learning holds remarkably little reference to mentoring. In time it will hopefully provide learning opportunities not only for pupils and teachers – but for mentors and trainee teachers.

There are a number of Internet sites that can support your growing interest in mentoring: (www.bath.ac.uk/Departments/Education/mentoring.html; globalmentor.org.uk; actionresearch.net; mentorsforum.net).

A list of key texts is given in the bibliography These should enable you to gain access to perspectives on mentoring that will enable you to understand and develop your own practice more fully.

The teaching profession is just beginning to see the potential that mentoring holds for improving schools and raising standards in class so that pupils can attain their potential and fulfil the high expectations that are rightly held for their development. Mentoring should not be a sideline activity to train pre-service and newly qualified teachers. It has a rightful place at the very heart of what a school is and does. Mentoring – educative mentoring – enables individuals to learn how to make effective transitions, to become more effective learners and thereby more effective teachers. Educative mentoring, unlike coaching, unlocks potential for everyone involved in a mentoring relationship. Lifelong mentoring does not mean keeping the same mentor for life but it does mean recognizing the value of having a mentor and seeking mentoring relationships when it is appropriate to do so; usually when you undergo periods of change.

At present mentoring is restricted by a lack of awareness about its potential for enabling personal and professional growth. Schools can offer mentoring to parents so they can assist their children in their learning. There could usefully be mentoring for new and would-be school governors or for experienced ones who wish to improve what they do. In a school like yours, why not promote mentoring for subject leaders and for headship? As a mentor you have a crucial role to play not only in your own development but in the future of our profession.

> . . . a considerable amount of research into mentoring has focused on descriptions and conceptions of what mentors say they do, and what they actually do. There is much less on what it is possible for mentors to do.
>
> (McIntyre *et al*, 1993)

By undertaking research you are redressing this imbalance.

Good practice for undertaking research to enhance your mentoring:

- You have much to contribute in increasing understanding of mentoring.
- You are likely to have the time and useful skills for research already.
- Assistance in carrying out research is widely available.
- You can submit your research for accreditation as a portfolio or assignment.
- You have a responsibility to your profession to undertake research.

# Part 2. THE PROCESS OF MENTORING

Chapter 5

# Observation of lessons

In this chapter we shall be considering the following aspects of mentoring:

- Organizing observation of lessons for your trainee.
- Observing lessons taken by your trainee.
- Identifying good practice in organizing lessons for your trainee.
- Identifying good practice in observation of lessons taken by your trainee.

## Organizing observation of lessons for your trainee

When you prepare your trainee to observe lessons consider three phases:

- Pre-observation discussion.
- Lesson observation.
- Post-observation discussion.

### Pre-observation discussion

There are two aspects to this pre-lesson discussion: your trainee needs to realize the overall aims of observing, and needs to learn to focus on predetermined objectives.

Aims and objectives are often confused but they are distinct. The aims of observation are to enable the trainee to see good practice in classroom teaching and to reflect on what has underpinned this good practice.

Both form the basis of assisting the trainee's development as an expert practitioner.

The objectives are to enable the trainee to identify the precise teaching techniques and strategies that enable learning and to try out these techniques and strategies in subsequent teaching.

Your trainee needs to understand that there are no 'fix-alls' and that teaching is an interaction, and at times an unpredictable interaction, between pupils and teachers. Above all your trainee needs to learn that the teacher is the single most important factor in motivating pupils. There is no ONE way to teach effectively because teaching is context bound, but some ways of teaching are likely to be more effective than others. Things may not work to plan for even the most experienced teacher but the trainee needs to see how to learn and grow from difficult experiences as well as good ones.

The objectives for the lesson observation might be identified entirely by the mentor at first but as training proceeds the trainee should become increasingly responsible for identifying a suitable range of foci. These complement training for development by augmenting the resource bank of strategies and techniques at the trainee's disposal to use within his or her own teaching. Why should trainees observe lessons?

- To ground their early idealism in reality while retaining a sense of 'wonder'.
- To demonstrate the expectations for work/behaviour that teachers set.
- To show that not only is survival possible but that teaching can be exciting too!
- To model ways of dealing with the kind of problems that often arise.
- To demonstrate that there are some problems where you 'just do your best'!
- To enable trainees to appreciate the link between pace and timing.
- To inspire trainees through creative, imaginative teaching.

## A phased approach for observation

Think about how it feels to be a trainee. You are suddenly in a situation where you sense familiar signals – after all you have been to school for many years yourself – but you know there are differences and, while you are keen to explore, you feel insecure!

When your trainee arrives in school it might be tempting to take him or her on a whistle-stop tour of the school, introduce them to everyone in sight and get them into lessons as quickly as possible . . . Your trainee needs to know so much at once and may ask you lots of questions. It is important to take the orientation process in a phased and rational manner, starting from building an understanding about how the school operates, enabling the trainee to understand what a typical school day feels like for a pupil and what a typical school day's teaching is like for a teacher in the school. Do include your trainee's views in decisions you make about a programme:

- Decide what your trainee needs to see.
- Remember that you will need to be flexible.
- Consider how you can phase observation for your trainee.
- Appreciate that the best-laid plans . . .

As an approximate guide to stepping observation useful foci might be the following:

Phase one: pre-teaching experience

- How a school operates.
- What are the school's expectations of a teacher?
- What are the school's expectations of a pupil?
- What is the ethos of the school?
- What constitutes a typical school day?
- Pupil pursuit – what the pupil experiences.
- Teacher pursuit – what the teacher experiences.
- Induction activities for pupils new to the school.

Phase two: early teaching experience

- Classroom management techniques.
- Basic question and answer techniques.
- Beginning and ending lessons appropriately.
- Managing different pupil grouping arrangements.
- Attending departmental meetings.

Phase three: increasing responsibility in teaching

- Linking teaching strategies to pupils' learning.
- Pupils with special educational needs (SEN) in mainstream class teaching.
- Parents evening.
- Observing pupil report writing.

Phase four: assessment time

Approaches to differentiation teaching.

Phase five: approaching QTS

Increasing creativity, imagination and experimentation in teaching.

There can and should be considerable flexibility in how you organize a phased approach for your trainees. Much will depend on their previous experience in schools.

## Post-observation discussion

The focus of your discussion with your trainee near the beginning of his or her training should be for you to discover the following:

- What are your trainee's preconceptions about teaching?
- What are the values that your trainee holds in relation to teaching?
- What does your trainee think constitutes good practice?
- What styles of teaching appear to be your trainee's preference?

Enriched with information winnowed from answers to these questions you are then in a position to fine-tune the mentoring programme that you and your trainee will be sharing. Your discussion is therefore diagnostic to your trainee's training needs as well as being informative by offering explanations as appropriate for your trainee. Over time your trainee needs to learn 'why' certain teaching situations arise and why certain activities are planned by a particular teacher at a particular time. Why are there no easy solutions? How can expert teachers come to manage their classes by pre-planning and by exploiting events that arise in the lesson. Trainees need to appreciate that training in how to research others' teaching is necessary if they are not to become too invasive in or too detached from their questioning of teachers.

The guidance given in McIntyre, Hagger and Burn is invaluable here and I have included appropriate sections from their handbook for mentors (1994) below.

Trainees should be given guidance so that they can identify with the teacher, imagining themselves in the teacher's shoes, facing up to the situations that confront the teacher and deciding how to deal with these. Thus in reviewing the lesson they should consider such questions as:

- What are the things that I'd find most difficult if I'd been the teacher in this situation?
- Would I have wanted to deal with them in the same way as the teacher did?

- If so, could I have done so?
- If not, how might I have been able to act?
- What would have been the likely outcomes of my preferred action?

Where it is clear during post-lesson discussion that trainees are overlooking fundamental aspects of teaching in their observation, and concentrating too much, for example, on classroom management, these should become the specific focus for subsequent observations by the trainee. Concurrently, they should also become the focus of the mentor's observations of the trainee because they may be overlooked.

## The kinds of questions to ask and not to ask

Trainees will take time to understand how teachers work and think. So often they will be tempted to accept rather than probe a teacher's motives and when they do probe, the questions they ask may be inappropriate. The training for a novice teacher needs to include a sensitization to teachers' needs, empowering them to know how to phrase enquiries. It is often not until a novice has asked an 'awkward' question or made an 'inappropriate' remark that their novice status is really apparent. The mentor must avoid reacting to questions and remarks as if they were being criticized by a fellow expert practitioner and should look on such contingencies as time to give more guidance in asking suitable questions and also to consider if there is validity in the trainee's comments – and if so, to use the observation as a basis for improving their practice.

Some of the information that trainees will gather during their observations around school are likely to frame much of the overall impression they have about teaching per se. They may not appreciate how different schools are unless they have previous teaching experience. They will frame their impressions on what they see and hear.

Trainees need to know that:

- teachers have feelings too and are also vulnerable to negative criticism;
- teachers do not automatically have a reason in their mind for their every action – their teaching has often become reflexive through experience – and certain situations bring certain teaching reflexes into play in the lesson;
- teachers are likely to become defensive if cross-examined directly;
- teachers (and mentors!) have the right and duty to say 'I don't know';
- teachers may find that an observer interferes with their relationship with pupils.

Above all, trainees need to learn that teaching is only effective if it results in learning – learning, that is, by pupils and by the teacher. They also need to appreciate that teachers can and should learn from their pupils. The management of this interaction is what distinguishes teaching from instructing, lecturing and training – it is what constitutes education. The objectives of post-lesson observation should be framed to increase the trainee's awareness of how learning and teaching are interrelated.

You set up a programme of phased observation for your trainee. You delegate responsibility to pupils and colleagues as appropriate and the time comes for your trainee to sit in lessons and take notes on what occurs. How do you approach this?

Much will depend on the individual trainee's attitude as well as experience to training.

---

## Illustration

Compare these two statements from trainee teachers: 'I've been observing lessons today – it was so *boring* just to watch, I can't wait to have a go' and 'I've been observing lessons today – it's such hard work – you really need to concentrate all the way through – I'm learning so much!'

Two novice teachers on the same PGCE course. Both are working with experienced mentors but there are marked differences that cannot be attributed to personality. The first trainee does not appreciate how observation will assist in teaching. The second trainee has had training in guided lesson observation.

Let us explore how the second trainee was prepared for lesson observation.

The mentor explained that she would be observing the lessons taught by this trainee in the weeks to come and she showed her the pro forma that she would be using in future. She set out the observation training in a series of stages. To begin with she set the trainee to record a description of what happened at 10-minute intervals through an hour-long lesson. By doing this she enabled the trainee to get an overview of what teaching a complete lesson involved. In the next week she asked the trainee to continue doing this but to make detailed notes about the way the beginning and ending of the lesson was managed in her class. In the following week she explained in general terms what the lesson would comprise in learning objectives for the pupils and asked the trainee to write down a plan for the lesson that she had observed. After the lesson the trainee and the mentor compared the plans – what the teacher had planned and what the trainee thought she had planned – and the mentor explained how and why the activities the trainee observed fitted into her plan – and where circumstances arose where she had diverted temporarily from her plan. By doing this she was reinforcing the concept that every lesson needs planning but that plans need to incorporate a certain flexibility if they are to be effective. Later she asked the trainee to observe a series of lessons in a class parallel to the one that the trainee was teaching. They compared and contrasted the plans and teaching techniques they had used. In particular they began to consider the role of the targeted question in addition to giving tests that enables a teacher to carry out informal assessment.

The other trainee was less fortunate. Although her mentor was skilled in many ways and she had selected carefully a range of classes by ability and age range and enlisted the assistance of other teachers to observe, her guidance too general: 'Go and sit in on these lessons and we'll talk later.' Indeed there was post-lesson discussion but the trainee's view was left at the idealistic stage. She saw that a teacher did A, B, C but not 'why' and it is exploring the 'why' that is the basis of effective guided observation. With so much to see and so little to focus on the trainee came away with the impression (largely unvoiced) that teaching must be easy and that she could succeed with the opportunity to teach. The lesson observations went on for a period of almost eight weeks because the mentor thought that the more the trainee observed the more she would learn about teaching. Not surprisingly the trainee floundered.

---

Discussion and observation should go hand in hand – throughout training. There is no cut-off point for lesson observation by the trainee and as the training continues the observation should become increasingly enquiring as a basis for the trainee to respond to the question 'How do I improve my practice as a teacher?' (Whitehead, 1989).

---

## Illustration

In this school trainees are regarded as being teachers from the very outset of their training programme. They are assigned to a pupil as a special needs support for a period of several weeks. When not engaged in teaching they are expected to attend designated lessons with the pupil to assist him or her in learning.

### Commentary

This is sound way of letting trainees experience the interaction between teaching and learning. Trainees are directly involved in the lesson because their help is needed frequently. They have to interpret what is coming from the teacher and frame it appropriately so that the pupil can learn. None of the pupils selected has severe educational needs in terms of behaviour or physical disability. The trainees supplement existing special needs provision in the school and have a choice of pupils with whom to work. Most have language problems, coming from a multi-ethnic background where English is their second or third language. The need for precise, clear explanations and instructions as well as the feeling of being a part of enabling a pupil to learn is a strong motivator for the trainees. The students welcome the additional assistance in a school environment that has now put ITT at the centre of its practice.

---

## Observing lessons taught by your trainee

In the introductory chapter to this handbook we looked at how trainee teachers' development appears to go through a series of stages. In your mentoring you might find it useful to think in terms of levels or dimensions rather than stages of novice teacher development. This is because it is quite possible for different levels to apply to different aspects of the trainee's teaching. There may be areas of 'early idealism' at same time as evidence of 'moving on' in other aspects of teaching. This can result from a host of factors relating to your trainee's personality and motivation for teaching as well as to the range and depth of any learning and teaching experience he or she has.

For the sake of clarity the phases set out below are presented in a linear fashion, but learning is very rarely linear and there may be apparent leaps in progression and then regression to an earlier stage as if the mentee checks out whether it was safe to have really made the initial leap at all. Broadly speaking progression will move through induction, consolidation and extension phases as the trainee becomes increasingly skilled and self-confident. The induction phase is where trainees gain an understanding of teachers' work in schools and begin to develop generic teaching skills. The consolidation phase builds on this developing awareness, strengthening competence, broadening and deepening what is being learnt, and the extension phase offers other opportunities to tailor training to move on from semi-dependence towards autonomy.

## Observation in the induction phase

When they first begin a course, trainees may be blissfully unaware of the expectations that the school, other teachers and pupils have of them. As awareness dawns panic may ensue and the 'survival stage' takes hold. At this stage the trainee will need close supervision as well as observation by the mentor who should be on hand to assist. It is very likely that the trainee will benefit most from working with just a small section of the class, perhaps a small group of pupils within a mentor's lesson. The mentor will plan the lesson and should explain to trainees well in advance what is to happen and exactly what role they should undertake with the group of pupils. Observation at this stage should be focused on seeing how well the trainee relates to pupils; is there a genuine wish to be involved; is there a caring manner; is the professional distance appropriate enough or is the trainee too ready to befriend, perhaps against the teacher?

Then what do I say?

We will look in more detail at how to give appropriate feedback to a trainee in the next chapter but in essence your feedback should be supportive and encouraging. It is useful to begin with praise 'I really liked the way you did A, B, C', and then move on to ask the trainee how he or she felt being within the lesson.

Once your trainee is showing signs of maturity in attitudes towards teaching and in appreciating the need for matching expectations to the capabilities of the pupils, through the experience of working with you taking the lead within the class, training can begin to move on to the second part of the induction phase. Gradually move your trainee into collaborative planning and team teaching, but retain a close presence.

## Sharing teaching

The shared teaching can begin with a small activity where the trainee has interaction with the entire class – ideally a small activity planned by the trainee geared to fit in to the lesson plan devised by the class teacher. It is probably wiser for the mentor to continue working with the trainee at this stage but it may be necessary, because of the distribution of classes, for the trainee to work in class with other members of the teaching staff. This can be a very useful experience, of course, as it gives the trainee an opportunity to share in different ways of planning and teaching a lesson. By preference, the mentor should team teach with the trainee on at least two occasions per week throughout the early weeks of training and more if the timetable allows.

As the trainee takes an increasing lead in planning and teaching the whole lesson it is only to be expected that difficulties will arise. Your trainee needs you or another mentor figure on hand to support his or her presence though it must be made clear that the mentor is not there to intervene but rather to support their teaching. If you should have to intervene, it must be in a way that does not undermine the trainee's presence and that offers a fallback policing potential. Mentoring focuses on classroom control, clarity of instructions and explanations, and effective planning.

## Observation in the consolidation phase

In this stage the trainee has begun to identify and deal with many of the outstanding problems that appeared so threatening in the induction phase and is taking on a period of extended teaching. During the period of orientation in the induction phase there may have been little attention paid to the question of assessment. The observation should now aim increasingly at enabling the trainee to meet the government's standards for qualified teacher status (QTS) within the areas of knowledge and understanding, and classroom management and planning, moving increasing into monitoring, assessment, recording, reporting and accountability, as well as looking to further professional development. We will look at all of these areas in later chapters.

## Sharing planning and teaching

For this section I have drawn from the chapter entitled 'Collaborative teaching' written by Katherine Burn in Wilkin (1992). I have amended Burn's universal definition of collaborative teaching to mean team teaching where mentor and trainee are both contributing to the planning and the teaching in an authoritative and collaborative way. I prefer the term 'shared teaching and planning' to describe the one that Burn uses in saying 'The intern (trainee teacher) has a clearly defined responsibility . . . which is deliberately targeted to help his or her learning.' Trainee teachers tend to overlook the amount of planning and preparation that marks out the practice of an expert practitioner because they do not see it written down. They cannot appreciate the internalized dialogue that is the real planning and can cut in at almost every conceivable occasion as the effective teacher plans both in advance of and during the course of teaching. If trainees could tune in to this dialogue they would be less likely to misread the way expert teachers work. As it is they can only see the teaching of lessons with little or no apparent forward planning, at least in a formal sense, and should they try to emulate this by thinking only on their feet we have a recipe for exhaustion, disillusionment and disaster! You must explain!

Collaborative teaching enables the trainee to see how a good teacher sets about planning in overview on a whole year level, using the departmental scheme of work, bringing planning down in overview to a termly and then a half-termly level, with core-plans on a weekly basis – fine-tuning to core/extension/contingency often as the lesson is getting under way. Your trainee needs awareness of this through being trained in more formal planning and preparation of lessons – the basis of expertise.

Even experienced teachers may have difficulty making what they do explicit. As a mentor I found difficulty understanding why a trainee could not see what was perfectly obvious to me; you need to plan to prepare for a lesson and you cannot leave preparation until the last moment or, worse still, until the lesson is under way. As experienced teachers, we have often spent many years in honing our teaching skills to such a point that the teaching does seem effortless and the explicit planning and preparation has become implicit expertise. Collaborative planning, if you are willing and able to share insight into why you choose particular teaching strategies and if your trainee is primed and confident enough to probe, will enable the implicit to become explicit so that it can be examined, quizzed and become implicit again!

Burn sets out the principles for good practice in collaborative planning and teaching:

- Both mentor and mentee must welcome collaborative teaching.
- Regular opportunities to collaborate are most useful.
- Roles in the collaboration need to be clearly defined beforehand.
- There needs to be opportunity for discussion during the lesson.
- Both mentor and mentee must be clear *why* they are collaborating.

Peer collaboration between trainee teachers can be a very useful strategy to build confidence and awareness at this phase of their development. By this stage, most trainees are establishing their presence in class so they are less likely to allow one or other peer to dominate in the lesson planning or delivery.

## Observation in the extension phase

Finally the trainee is coming into his or her own and has developed sufficient expertise to teach some good lessons fairly regularly. In many ways this is the most exciting phase for mentors because you can see the fruits of many hours of training, encouragement, modelling and education coming to fruition. A colleague emerges from a trainee! Using this emerging expertise is one of the most challenging and rewarding things a mentor can do and has potential for professional development for both parties. The trainee no longer needs supervision in class and is becoming autonomous. It is tempting to allow the development just to ease along. It makes more 'mentoring sense' to seize this opportunity to work creatively together; to look at the scheme of work and see how it can be realized in individual lessons drawing on the imagination of two teachers. There are lamentably few opportunities in full-time teaching for such collaboration in expert planning and mastery of learning-through-teaching in a collaborative way. Trainees can bring knowledge and experience from a training course where they have met the very latest teaching techniques, and they already know about research relevant to the classroom.

Your trainee also has recent experience of working in more than one school and may well bring expertise from a previous professional post. Now is the time to draw on this new expertise and for you to learn new teaching techniques and approaches – perhaps using computers in a more wide-reaching and appropriate way to bring opportunities for your pupils to learn more and learn better than they could before. Mentors in the Bath University PGCE Partnership often comment on how this final phase of training repays some of the time and effort that they have invested beyond the timetabled mentoring hours. It is a time for sharing expertise, for reflection on practice, for all.

Good practice in organizing lessons for your trainee:

- Phase your programme of observations so that you don't overwhelm.
- Orientate your trainee to your school as well as to teaching.
- Manage the process of discussion about observation in a logical way.
- Start from your trainee's needs as well as from what they ought to know!

Good practice in observation of lessons taken by your trainee:

- Give your trainee insights into how you plan and prepare for teaching.
- Reflect on what your trainee needs to do to plan and prepare.
- Give assistance in planning and preparation, but don't do it *for* your trainee.
- Build observation from team teaching and sharing collaborative planning.

Chapter 6

# Giving feedback to your mentee

In this chapter we shall be considering the following aspects of mentoring:

- Getting the context right for giving feedback to your trainee.
- Giving feedback after the first teaching experience.
- Giving your feedback in steps to enhance your trainee's development.
- Identifying good practice in giving feedback to your trainee.

## Getting the context right for giving feedback to your trainee

Instead of just providing you with a list of guidelines for giving effective feedback, we should start by considering the answers to a number of questions about the context:

- What is your trainee thinking after teaching a lesson that you have observed?
- What is he or she expecting you to say?
- Will the thoughts and expectations differ according to the background, perhaps a previous career, that your trainee has experienced?
- How will your trainee's expectations change as the training progresses?
- How will you empower the trainee to help you improve your teaching too?

So often 'good advice' in mentoring assumes a rather static model where giving feedback is concerned. It is true there are certain commonalties in good practice no matter what the context:

- Being honest.
- Empathizing with your trainee.
- Keeping to the facts rather than to opinions.

It is also important to give both support and challenge – but how much and when?

Let us start by considering how a trainee teacher who has just taught his or her very first lesson in the course might be approaching a feedback session. There are few teachers who do not recall something of that first lesson: what they had to teach; how old the pupils were; the rush of adrenaline; knowing that all eyes are focused on you – or seem to be! The excitement, the vulnerability, the relief when it was over?

In the following illustration I give an account of my first experience of teaching. As you read it think about the differences and similarities in your own experience and in that of your trainee who has or is about to teach that very first session.

---

## Illustration

When I recall my very first lesson, which I guess lasted about 10 minutes, it seemed to go on for ages. I was teaching a fable by La Fontaine to a group of nine-year-olds. I had made visual aids: large posters of the main characters. I can still remember the sunshine coming through the library window to the left of where I was teaching. I can remember the group of children, although their individual faces have melted into oblivion because it is many, many years ago. I remember the smile on my mentor's face! I remember his words of encouragement. Constructive criticism followed later – the first message I took from my mentor was yes – I had *taught*. It was so exciting – especially as I was *never* going to be a teacher. I was far too timid – at least I thought so because my school reports said so – I could never be a teacher...

## Commentary

Recalling the first experience of teaching is rather like those major events where the world seems to slow down – the death of John F Kennedy, the death of Princess Diana and when midnight chimed the world into the new millennium! Special moments when the mind is focused, the senses speak and the emotions run high.

Consider for a moment:

- Your trainee may be like me – very unsure that teaching is the right career.
- Your trainee may be a 'born' teacher, for whom no other options appeal.
- Your trainee may have left a career in banking perhaps, or running a home.

One thing is sure – no matter what their background, trainees need mentoring! Some trainees may be open and childlike when they have been teaching, some less so; more reserved, self critical, defensive, some almost nonchalant . . . No matter how your trainee reacts, he or she needs feedback from you.

---

## Giving feedback after the first teaching experience

The first thing that you must do is to acknowledge that an important event has just occurred. It is a privilege for you to be present at such a momentous occasion – so celebrate that! Immediately after the teaching has taken place, after these first few steps in a new profession, thank the trainee for the opportunity to be there and express your delight that he or she is entering teaching – and *smile*! There is no need for a lengthy mentoring session immediately afterwards – by all means say that you will meet up later to talk over how it went – but give immediate praise: just one or two words until the mentoring takes place.

At this first vital mentoring session you are setting the ground rules for your mentoring relationship. First impressions of your mentoring are vital. Keep to the point – say 'I liked the way you . . .' 'The pupils . . .'. Be approachable, honest and supportive. Ask the trainee how it felt to teach and let him or her discuss their emotions if they need to – as well as any events. You may decide to ask exploratory questions but do so gently and listen to your trainee's reply before you respond.

- What do you think went well? What else? (Three things, perhaps.)
- Yes, I thought so too! What about . . .?
- Why do you think it went well?
- What do you think didn't go quite as you planned?

- What might you do differently next time?
- Shall we have a think about that when we meet on . . .?
- I also liked how you . . . what do you think? How did you feel when . . .?

Once the heat of the moment subsides, they may say *'I think I could have . . .'* Excellent! The beginnings of reflection on their actions (Schön, 1984) are taking shape. Try to agree at least in part and before suggesting another approach: *'What you did worked well. As we work together we will be looking at other ways of doing things in our lesson – when you might like to try . . . we could think about . . .'*

The trainee has just taught his or her first whole lesson. An entire hour and your trainee is tired and reasonably satisfied – relieved and confident. *'Thank you – I enjoyed your lesson. I'll give you a chance to catch your breath. We'll meet at . . . so we can talk over how it went. Well done!'*

The first proper mentoring session approaches. The trainee wants to know not only how it went but how he or she measures up as a teacher – already! Even if your trainees do not ask you directly how they are doing, that question is likely to be foremost in their mind most of the time. It *is* too early to say if a trainee will make an excellent or even a competent teacher at this stage but make an effort to stress even the small things you have observed that suggest to you that he or she will succeed in teaching.

The feedback session should be as soon as possible after the lesson and in a private place, somewhere quiet and confidential and not the middle of the staffroom! Begin with reassurance and remind your mentee about why there are mentoring sessions. 'We can share a regular mentor meeting at . . . in . . . (room) where we can discuss how your teaching is going, looking at the good and the not-so-good points and how we will help you teach well.'

Try asking how the trainee felt about that first lesson because this is a diagnostic mentoring session. You need to know how realistic and open-minded your trainee is:

- Encourage your trainee to tell you three things that went well.
- Praise these aspects so you can begin and end the session on a positive note.
- If things went badly or if your trainee is feeling disheartened, be gentle – a hurting trainee is not a receptive trainee! Be supportive but honest.

'X went really well because you . . .' In your feedback demonstrate that you want to identify more than just a list of good and not-so-good points, but you do want to explore the reasons for these. Your trainee is perhaps in the haze of the early idealism stage, where the focus in his or her mind will be on performance and little, if at all, on monitoring the interaction between the act of teaching and any learning by the pupils. No matter how good or how poor a teaching session has been, give praise where appropriate, set targets and work on ways of meeting these targets through action!

## Giving your feedback in steps to enhance your trainee's development

The first lesson has come and gone. The trainee is taking a few lessons each week now and you are observing at the very least once per week. It is Friday afternoon and the class, a year nine, has been torturing your trainee. He or she is upset. Internal voices in your trainee's head are saying unkind things! 'You'll never make a teacher! You can't control the kids – it's their fault! It's Friday afternoon! The kids won't behave!'

The 'survival' or rather the 'lack of survival' stage is dawning. What do you do? Somehow you have to reassure your trainee, you have to convey that pupils are teachable even on Friday afternoons and that there are ways of doing this – all without making it seem as if you know everything and the trainee knows nothing, which is often how trainees feel at this stage of their development. To make matters worse there is no mentoring session or timetabled feedback time for a few days; after the weekend.

What must you do? First you must reassure – and be realistic. You don't want to be cast in the role of policeman or woman every time a class plays up your trainee *but* . . . he or she desperately needs your support so you must make time *now*.

Before the trainee leaves the school for the weekend you need a brief mentoring session. In-depth analysis can wait for next week but your trainee needs you to give a balance to what has just happened and in your creativity as a mentor you must offer this. Give only appropriate praise – any more can seem patronizing and false. Praise what went well – no matter how small – and *then* ask the trainee about what went badly. If necessary say 'I realize that you are not happy with how this lesson went. We all have times like this and we have to look at what didn't go so well and then work together on getting things better next time!'

The 'we' is important here. You are reducing a sense of isolation which your trainee is probably feeling and very likely some of the pent up anger too. Make it clear that one poor lesson is not the end! If possible talk encouragingly about other lessons taught during the week – and assure the trainee that you will look at this particular lesson in detail in your timetabled mentoring session the following week. Set the trainee a task (nothing counteracts fear like facts) and agree that the next mentoring session will focus on a list of just three main points. From your observation notes pick out only three that need attention and ask the trainee to have some thoughts about the possible reasons for these. Finally, agree with your trainee that you will work on these together, next week! If your trainee seems still very distressed and you have the opportunity to make it, a quick telephone call over the weekend to reassure and comfort can make a world of difference.

The following week when emotions have quietened down and your trainee has had time and space to reflect (and so have you) you can set about analysing what happened. Again, it is important to be honest – saving your trainee's feelings at this stage may inhibit later progress. But give feedback in a spirit of compassion and care.

## Next step!

It is a scheduled mentoring session. Your trainee arrives and wants to discuss a lesson plan. You see that the objectives revolve around class control and you know this isn't 'teaching' but your trainee is insistent that you give tips for effective teaching. 'Tell me what to do so it will work.'

What should you do? Perhaps the first thing to do is to ask yourself why your trainee is posing this demand. It may well be that your trainee is already aware of the need for sound classroom management so that learning can take place, but there could be other motives driving the question. Is it because he or she is genuinely unaware that there should be more in the lesson plan relating to learning than classroom control? Is it that he or she is becoming frightened and classroom control is the dominating feature of any lesson at present – a need to survive rather than ensuring that pupils learn? Is it that on the whole it is easier to think about the 'mechanics' of teaching – the classroom – than grappling with the idea of effective teaching bringing effective learning? Is it that you have been giving only 'tips for teaching' in your mentoring?

How you interpret the demand will have a substantial effect upon how you frame the response you give but the content of the mentoring will have to relate to enabling the trainee to see:

- There are solutions to most difficulties.
- There are no hard-and-fast rules for solving problems.
- Classroom control, although essential to effective teaching, does not replace it.

So begin by resisting any attempt to get you to 'sort it all out' for the trainee. Instead, look for ways of reaching solutions between you and your trainee for the more outstanding problems. You can begin through targeted questioning: 'When you watched lesson X how did teacher Y deal with . . .? Could that work here? How would you have done Z . . .? Have you tried it?'

It is important to move the trainee on to focusing on what the pupils should be doing while he or she is teaching a particular point or when there is an activity for them. Thorough questioning and reflection with the trainee, challenge and explore possible strategies and solutions, drawing on his or her experience so far from teaching and from observation.

It will be tempting to leap in and rescue the trainee but rescuing is less successful than empowering. What does work is to ask the trainee to consider how the lesson planned will be experienced from the standpoint of the pupil. Are there long periods of inactivity when it could get boring? Are there sufficient different types of activities to motivate and interest? Work towards a plan that is composed by the trainee with your assistance; a plan that will be satisfying for pupils and for the teacher alike. If trainees feel insecure, look for 'safer' activities so that they experience success.

## Stepping out

Your trainee has been teaching several lessons that demonstrate an ability to deal with most of the problems that arise in the classroom. Minor disruption is addressed swiftly and effectively and more major problems are being handled reasonably well even if the trainee is seeking assistance. The lesson plans are revealing 'safe' activities and, although the children are occupied, they are not being stretched. The trainee is feeling confident. Where do you go from here? Your trainee is not asking many questions but is looking at the mentoring as a means of gaining your approval for what has been planned. How do you retain the trainee's self-confidence and move him or her on to better practice? Do you know how to improve on the plan if it differs from your way of setting up activities? How do you advise without running the risk of cloning? How do you encourage the trainee to increase the range and scope of learning opportunities? Time to start considering differentiation and creativity in more depth!

## Stepping out into a run!

Time also to ensure your trainee understands the significance of assessment:

- Is your trainee aware that assessment means more than giving tests?
- Does he or she see you assessing and monitoring pupils' progress informally (watching body language, keeping an eye, targeted questioning) as well as formally (testing under examination conditions)?
- Does your trainee yet realize that by assessing and monitoring pupils' progress planning can more effectively meet individuals' needs in class?
- Does your trainee know how to record monitoring and assessment using attainment target levels from the National Curriculum? And is he or she doing so?

We will be looking at the subject of assessment in chapter 19 but since it is one of the areas of classroom teacher practice that meets the strongest criticism from OFSTED inspectors, we should at least mention it in relation to giving appropriate feedback to a trainee. In your feedback to your trainee you need to make it clear that:

- Formative assessment should be used in a systematic and sustained way.
- It can significantly improve degrees of learning in the classroom.
- Planning to teach without planning how you will assess the effectiveness of your teaching is not effective practice. Assessment needs planning too!

So, if your trainee is teaching to the middle range of every class, and pupils are barely being challenged and engaged in learning beyond this, look for evidence of planning that uses the outcomes of assessment to inform and direct it. Challenge your trainee to tell you how each pupil is progressing in a particular class in a particular attainment target, perhaps. Stress at every stage that teaching without monitoring, assessment, recording and reporting, and using these to inform planning, is inadequate.

## One step back?

You are faced with a dilemma! Your trainee has just returned from teaching in a second school placement and he or she is not following your procedures! After a lesson you are to give feedback and you realize that you do not agree with some of the teaching habits that have been acquired recently. When you ask why the trainee does X or Y he or she says 'because my mentor said I must do that – s/he was surprised that I didn't know that already.' You feel taken aback and maybe not a little annoyed about this intrusion. You could say 'Well, we have agreed as a department that we will do Z' or 'That's a great idea, we will have to consider doing that here – let's think about it and maybe raise it at the next departmental meeting', or try to ignore that anything has changed, or consider it is not worth mentioning. The first question to ask yourself is 'Does this new strategy result in better learning or more effective practice for this trainee and perhaps for me?' If so, consider taking it on board – mentors are learners as well as teachers! If not, you can accept that this is part of the individuality of the trainee emerging – and if the strategy impedes learning, you should counter it . . .

You may not find this easy – but try to make it clear that you value the trainee without running down the practice of another mentor or school. Look for positive ways of raising your trainee's awareness about the negative implications of this habit!

## Mastering the quick-step!

It is the final stage of the initial training course. Your trainee has now secured a teaching post for next year and is teaching really well. You are very busy and tired and could do with some time to catch up! It is tempting to abandon formal mentoring sessions now, to get your trainee to take lessons for you – after all it *is* good experience for him/her. The trainee is happy to oblige. Do you just hand over some of your classes? There is something to be said for doing so. You have doubtless given far more than the allotted time to your trainee over the course of the year in terms of formal and informal mentoring. You have earned some time and your trainee will benefit from extending the range and scope of teaching. But there is an ethical question to address here. Your trainee is not a qualified teacher but at the threshold of a new career, and has probably met only the minimum standards as yet in terms of achieving qualified teacher status. It would be fairer, for both you and the trainee, to identify those areas of the Career Entry Profile (you can find out more in chapter 14) that relate to further professional development and to give opportunities for the trainee to have experience in teaching to develop those areas. In return, a certain amount of 'paying back' is only fair as long as your trainee will benefit too. Your trainee is not a 'cover' teacher! Trainees should continue to receive training to consolidate their skills. You can assist by widening their experience of teaching, or arranging for visits to neighbouring schools, perhaps to increase their repertoire of teaching strategies. They might also be happy to contribute to staff ICT training.

By the end of the initial teacher education course you will have been called upon to give feedback in a formal as well as an informal way while your mentee has been developing through broadly typical stages of early idealism, seeking to survive, dealing with problems, plateauing and moving on. In the feedback sessions you and your mentee will come with agendas. How you meet the needs and demands of your trainee and simultaneously those of your understanding of the ITE course that you are delivering will evidence your professionalism as a teacher educator. Giving feedback should never be formulaic; there is no one 'right' way since feedback, like teaching, depends on interactions between teacher and learner.

## Identifying 'generic' good practice

It is, however, possible and useful to look at generic guidelines for 'good practice' in giving feedback and these are set out below.

Giving feedback is arguably the most skilful part of the mentor's role. It is dependent upon accurate and detailed observation of practice as well as an understanding of your trainee as an individual learner. The dynamics of giving feedback are many and complex. The trainee is likely to feel under considerable pressure to conform to what he or she sees as your ideal in a trainee and may play safe rather than risk experimenting. On the other hand your trainee may have such a strong idealised image of themselves as teacher based on their impressions of 'good' practitioners that it is difficult to enable them to develop a teacher identity of their own. Hawkey (1996) points out usefully that beginning teachers need to be able to explore their early images of themselves as teacher – and they need the assistance of mentors to do this if they are to develop properly. They need to see that teaching is determined largely by the context in which it takes place. Good practice may not easily transfer from school to school. With help from you they can begin to understand 'generic' good practice but cloning through compliance without question is a constant danger. Reflective practice in giving feedback means exploring alternatives . . . and enabling your mentee to see as well as seeing for yourself!

## What do trainee teachers expect?

Asked what qualities a trainee expects to see in a mentor who is giving feedback, a group of PGCE students at Bath University raised points of which this is a summary:

- Being honest, approachable, friendly and a good listener.
- Giving (and taking) constructive criticism about teaching.
- Being able to pinpoint areas in your teaching where you feel weak.
- Working with you to improve those areas of weakness.
- Being prepared to raise and discuss those areas that you feel are difficult to discuss and delicate but are vital to your development.

Structuring your feedback is essential and thinking about it in terms of a series of steps may be helpful to your future approach. There are some ground rules that should help you to give effective feedback.

Start each debriefing session with open questions such as:

- What did you think of the lesson?
- How do you feel it went?
- What do you think went well?
- Why do you think this worked?

There is a saying in business circles that in every mentoring session you should ask five 'why' questions. Where your trainee needs to be challenged, just try it out!

Questions to ask yourself:

- How can I improve the quality of feedback I give to my trainee?
- How can I ensure that my trainee gets appropriate feedback from other teachers?
- How can I encourage my trainee to give me appropriate feedback so I can learn too?

Good practice in giving feedback to your trainee:

- Be positive in your feedback.
- Break down your feedback into sections relating to different aspects that you have observed in your trainee's work.
- Follow a broad set pattern for giving feedback which has a balance between listening and talking, supporting and challenging,
- At the end of the feedback session, encourage the trainee to summarize what has been agreed by you both as the target(s) for development.

Chapter 7

# Planning lessons with your mentee

In this chapter we shall be considering the following aspects of mentoring:

- Understanding how your trainee approaches lesson planning.
- Dovetailing your mentoring to your trainee's development.
- Enabling your trainee to construct a good lesson plan.
- Identifying good practice in planning with your trainee.

## Understanding how your trainee approaches lesson planning

There are likely to be quite dramatic changes in the way trainee teachers approach planning lessons as their initial teacher training progresses, especially if there has been little or no previous teaching experience. There are some similarities to the scenario where a teenager wants to learn to drive a car! At first there is the passion to learn ('I've been waiting for *years* to drive!'), often an underestimation of what is involved ('I'll only need a couple of lessons and I'm sure to pass first time!'), and driving is seen as a means to independence ('you won't need to come and ferry me round any more') rather than as a fairly laborious process to master.

This passion and underestimation typifies what Furlong and Maynard (1995) term the 'Early Idealism' stage. As driving lessons get under way there are often moments of sheer panic – 'There are just too many things to remember at once!' – and that is how teaching can seem too. This parallels the 'survival' stage for trainee teachers as they often attempt to 'get through' lessons rather than being able to exploit the creative potential. The focus is often on keeping the pupils occupied, controlled and quiet and there is little appreciation of the interplay of teaching and learning. Just as the car sometimes gets the blame for not responding to instructions, the pupils often get the blame for not learning. It may be fair but it is more likely to be poor technique. As yet there may be little sign of the mastery that good teachers develop to orchestrate their classes, to achieve a harmonious balance, through the selection and planning of activities that maximize each individual's potential to participate and learn.

The next stage or dimension (note these are not *necessarily* linear stages) in trainees' development sets in as they start dealing with difficulties, mastering class control and planning 'safe' activities that occupy rather than stretch. Taking the driving analogy one step further this is like learning how to reverse a car round a corner, getting road positioning right but driving in a rather decontextualized and reactive manner rather than a planned, proactive one. Once classroom control is mastered and most pupils are learning at least something constructive in a lesson, we may perceive the overassurance and sluggishness of the 'plateau' stage. This approximates to driving only in daylight and only on A roads. Both the trainee and the driver can be said to plan, but they are planning in a too restricted way. If they are to 'motor', they need to gain confidence and plan proactively, creatively and imaginatively and bring enjoyment and development of skills to making their journey from A to B. In class this parallels with developing the ability to differentiate work so that all learners can be supported and assisted to achieve learning potentials.

Finally we get to rally motor-cross, and maybe one day to the Grand Prix – the excitement of teaching well! The final stage identified by Furlong and Maynard, the 'moving on', is at the threshold of advanced driving. Now you are motoring on motorways, at night, in snow and ice: now you welcome challenge, it's really fun!

What tends to happen at the earliest stage of teaching is that novices see experts teach and simply do not appreciate the expertise in terms of planning that underpins the expertise in terms of practice. It looks easy! The art of the mentor begins in preserving the trainee's sense of wonder and making it clear that good teaching demands really sound planning. This is not a sexy message! The expert driver had to learn how to turn on the ignition, change gear and manoeuvre round corners before mastering the screech of the skidpan and the scream of the racing circuit. Getting the balance right; losing sleep over the traffic jams in driving, losing sleep over the trivia in teaching and it becomes a chore. You may not be able to avoid the trivia, the form filling, the endless marking when you'd rather be in class alongside the pupils, but you can manage what you do so the trivia descends to its proper place and does not destroy your vision of what you and the pupils can and should achieve together.

Trainees usually come with a host of ideas, full of preconceptions and misconceptions. They often have a very clear impression of the kind of teacher they want to become but they can loose this inner drive when the going gets tough and, like mind-numbed clones, they just 'do as my mentor does'. Planning needs to allow for their personality and individuality to shine through while meeting their needs as vulnerable trainees in a tough environment. The mentor, like the driving instructor, needs dual controls at first. Driving instructors do not do the driving for the trainee (that's why there is no steering wheel in a driving-school car) but they keep trainees out of danger (brakes) and avoid too much damage to the vehicle (clutch) and other drivers (colleagues) and pedestrians (pupils). Dual controls and being 'on hand' are helpful but only if the mentor like the driving instructor has the foresight to see when trouble is coming and reacts!

## Dovetailing your mentoring to your trainee's development

The experience of teaching that trainees bring with them will vary enormously. You may find yourself with the absolute beginner, not that long out of school, fresh from studying for a degree and with a vision of teaching from the 'front side of the teacher's desk'. On the other hand your trainee may already be among the band of seasoned teachers whose motive for ITT is to acquire professional qualifications.

Get to know your trainees before finally deciding on classes to allocate them. Match the available classes in the best constellation you can to their needs and past teaching experience. Set up an individual education plan for each trainee.

Let us take a look again at the approximate stages of a trainee teacher's development (and you can find out more about working with these stages throughout this book). How does lesson planning evolve and how can you help it along?

### *Stage one: early idealism*

Your trainee needs to see sample models of lesson planning by you, the mentor. Without this, your trainee cannot be expected to know what to aim for. If you normally just plan something as you go along, on the way to school, as you open the classroom door, things are going to have to change! You can hardly write a teaching report, decrying your trainee's inability to plan properly, if you don't appear to do it properly yourself! What is useful for a trainee is for you to talk through your planning from the framework of a ready lesson plan pro forma that has been recommended for use during your trainee's initial teacher training course. It can be too confusing for a trainee to find that the course recommends using one pro forma and you insist on using another developed in-house. It is helpful to transfer this on to a computer so it can be completed, stored, saved and reproduced at will and subsequently amended as the need arises. The conversation you have with your trainee might run along the lines of 'This is the class I am planning to teach. What do you think are the things I have to bear in mind as I plan? Let's see if we can identify them in the lesson plan I prepare.'

This strategy can be an excellent means of furthering the professional development of your trainee and yourself. We also tend to 'plateau' in our own teaching. This is a timely reminder to consider all the points we need to bear in mind to ensure continuity and progression between lessons, and to make these explicit in planning is good practice.

### *Stage two: survival*

Co-planning comes into its own. Your trainee needs to see good practice modelled and begins to take responsibility for planning what will take place in the lesson in a more structured way. But without your close questioning to check the viability of the plan and your ensuring the preparation of resources ahead of the lesson your trainee has problems in store! Again you needs to demonstrate how and explore why (the 'why' is especially crucial) certain activities are included in this lesson plan.

Without this what does your trainee do? The simple answer is that he or she will often flounder around, left only with the following possibilities:

- Hope to have thought things out sufficiently by themself.
- Copy what you do and hope to pull it off effectively.
- Watch the clock and pray for any release from purgatory.
- A combination of all three!

It is unlikely that a trainee will realize, at this point, the dynamic that must exist between the plan and the teaching context, between what he or she does and what pupils should do. It is at this stage that most trainees will encounter problems and need most support from their mentors on a personal as well as on a professional level. Imagine that your trainee plans a lesson, it starts well but midway there is minor disruption (or worse – all hell breaks out). Your trainee has managed to visualize himself or herself into the early part of the lesson but being unable to project beyond that loses an image of how to 'hold the class' by keeping the right dynamic and interaction with the pupils going.

The trainee needs to appreciate the dynamic interconnection of teaching and learning. Trainees 'see' the image that they have created internally of themselves as an expert teacher put on the line. They may not see a failure to plan an effective lesson in a flexible way.

Because trainees are at the survival stage they are often clinging frantically to the detail of what their plan says *they* must do – the emphasis is on *their* performance rather than on how *their* teaching should evoke learning by *others*. This explains the significance of your asking 'why?' and probing gently until a response emerges. When you review a lesson plan created by your trainee:

- Applaud the fact that planning has occurred.
- Point out aspects (there must be some) that you know are useful.
- Ask your trainees to explain how they have planned and why they have done so like this.
- Challenge your trainees to tell you what the pupils are doing during activities.
- Look for ways to advise for smoother transitions between activities.
- Co-explore possibilities to extend learning so they can experience success.

It is important to bear in mind that this is a lesson to be taught by a novice. Where more experienced teachers can take risks knowing that they can 'bring round' a less-than-successful activity by reading their class and adjusting as they teach, few novices will have mastered this technique. Planning for lessons needs to be supportive both for the trainee and pupils. You should ensure that the plan is within not only the class's capabilities but also the trainee's capabilities and you simultaneously need to check that it is suitably ambitious in scope. Too safe and there is little room to grow accustomed to the very flexibility that good teaching involves; too risky and your trainee may flounder. Getting the balance right for you and for your trainee is of paramount importance for everyone; not least the pupils who are right on the receiving end of poor planning!

### *Stage three: dealing with difficulties*

The key to successful mentoring at this stage is to enable your trainee to plan effectively so problems rarely arise, or at least the more avoidable ones do not.

What seems to evade even the most intelligent trainees is the realization of *how* they have directly contributed to the problems that arose in the classroom. At first they may well try to go it alone and sort out every problem by themselves (often with disastrous effect) or they will try to hand problems over to you as minder:

- You need to remind your trainee that there is a school reward system.
- You need to point out that using this is preferable to the sanction system.
- You need to demonstrate how to operate both systems in unison.
- You need to advise when it is and isn't appropriate to use either system.

Your trainee is only just beginning to value your school's support systems. You may have pointed out the rubric in the staff handbook and you may have spoken about need to use rewards and sanctions, but until your trainee appreciates how you use these with different groups of pupils and in different teaching contexts, talk of rewards and sanctions is no more than rhetoric. To build an appropriate professional 'presence' as a teacher, your trainee needs to appreciate the value and not just cost.

## *Stage four: hitting a plateau*

By now, your trainees are dealing with minor disruptions and planning to avert major crises but their teaching often fails to avert longer-term, more insidious problems, most notably under-achievement among quieter pupils. Although your trainee is teaching fairly effectively to the middle of the class's ability range, and personal as well as professional objectives set out in lesson plans are being largely met, this is not yet 'creative teaching'. The temptation here for you both is to ease back. Trainees are often near the end of their initial training by this stage and have largely met the minimum entry requirements for gaining qualified teacher status. The lesson plans may incorporate opportunities to demonstrate skill in achieving a few remaining standards for QTS but increasingly your trainee is keen to be left to his or her own devices. However, your mentoring is far from complete!

Your trainee is in danger of doing just enough to keep most things ticking over in the classroom but missing opportunities to improve teaching and to extend the pupils' learning.

There is no longer a problem with classroom discipline, if there has been one in the past, but the lessons lack a certain professional sparkle that would make them memorable. This is the doldrums of the teaching journey when just enough might seem sufficient. In your mentoring you can assist your trainee by:

- monitoring the detail of the lesson planning by your trainee;
- ensuring that all learning outcomes are appropriate to extend pupils' performance;
- checking that lessons are not being mindlessly duplicated across year groups;
- reminding your trainee that good teachers always seek to improve their practice.

How you interest your trainees at this stage is likely to affect their professional development for the whole of their career. If you step back and aspire to anything less than really good practice then your trainee is likely to do so as well. Model in your own professional practice that you expect to use every opportunity to become a better teacher.

### *Stage five: moving on*

A really exciting stage both in learning to teach and in being a school-based mentor! Now your mentoring should consist of more challenge than previously if your trainee is to move beyond doing 'enough' and move on to be an expert practitioner. Less overt personal support and more professional challenge is required. How do you manage this transformation? In some ways you can't, but you can provide opportunities for this to occur by:

- ensuring that your trainee teaches more challenging classes;
- stressing the need for continuity and progression in pupils' learning;
- underlining the role of assessment in review and in forward planning.

Your trainees need to move beyond the 'safe' classes, the nursery slopes where they started to experience what it will like to teach after the first stage of ITE is complete. Often trainees will have secured a teaching post and may feel that they are marking time until it is time to move on physically. The challenge facing you as the mentor is to demonstrate by sharing that good teachers remain learners throughout their career and no teacher can know everything about teaching. Encourage them to look at their own teaching in a critical way, to set their own goals and action plans. It may be a useful time for trainees to consolidate a resource file of overhead transparencies, games and worksheets for areas of the syllabus that are likely to be common to any syllabus – all, of course, thoroughly piloted in advance. There is probably time now to hone ICT skills and to concentrate on effective differentiation to meet the learning needs of all pupils. Above all, this is the time to move from competence to mastery, to go beyond entry level requirement into the profession of teaching. This may well be a challenging time for you as the mentor too, as your trainees reject some of your teaching strategies that you hold dear and which work well for you. At last this is the platform for sharing expertise. Your trainees are coming of age, developing a 'style' in their teaching and seeking excellence.

You should remember that unfortunately not all trainees are capable of reaching the final stage of ITE development before they leave the first stage of their initial training. You have only partial responsibility for their growth. Some have areas of basic training that will dominate the scene right until the end of the ITE course. It is important for you to realize, however, that all trainees bring potential strengths and your role is to enable them to learn to plan so that these strengths are maximized in class.

## Enabling your trainee to construct a good lesson plan

Having talked in quite general terms about the changing face of mentoring in relation to your trainee's learning, and the significance of planning, now let us look in detail at one central aspect of the planning process; the 'well-tempered' lesson plan.

---

### Illustration to share with your trainee (adapted from the *Bath University PGCE Handbook*, 1999–2000: 24)

Preparing a lesson involves knowledge of the subject matter and of the pupils, and knowledge of appropriate aids, techniques and methods.

Lessons should be prepared far enough in advance to give time for the collection of necessary materials and illustrations. Experiments, demonstrations and any equipment to be used should be tried out beforehand. The resident teacher is not your personal laboratory technician or librarian.

Your notes should show the orderly development of the lesson in some detail including, for example, key questions to be asked, the use to be made of textbooks, the board and other visual aids, and the planning of the pupils' practical and written work.

You should define the precise aim and purpose of the lesson as distinct from its subject and title, and as distinct from the general aim of teaching the subject as a whole. In other words, what do you assume that the children know or can do before you begin? And what do you intend that they should know, or be able to do, after you have finished? The differences between these are your intended 'learning outcomes'. Where appropriate these learning outcomes should be referenced to the National Curriculum.

The subject matter of the lesson should be clearly distinguished from the methods by which it is to be taught.

Pupils' activities and opportunities for cooperation should be planned and indicated in the notes.

The work to be accomplished should be summarized (preferably by the pupils or at least with their cooperation) at convenient intervals and/or at the end of a lesson. Such summarizing should be indicated in the notes.

As soon as convenient, an opportunity should be taken to review the success or otherwise of the lessons, under the heading 'comments and criticisms" or 'evaluation'. These comments should refer to the planned learning outcomes.

---

## Would you know of a good lesson plan . . .?

Think of the lesson plan in four sections:

- Section A: overview of the teaching/learning experience.
- Section B: review of the teaching/learning experience.
- Section C: detailed account of the core activities planned.
- Section D: overview of the extension activities planned.

The lesson plan is designed to fulfil a number of aims:

- It shows the trainee the interaction of teaching and learning.
- It reinforces the need to consider continuity and progression.
- It demonstrates the complexity of elements in effective teaching.
- It is an informative basis for future effective lesson planning.

It is also designed to meet several objectives:

- It provides a checklist for the trainee to use in preparing a lesson.
- It provides a visual record of what was planned against evaluation.
- It provides a focus for an observer to evaluate during the lesson.
- It provides a focus for discussion in the mentor feedback session.

## Ingredients of an effective lesson plan

As we gain experience and expertise in teaching we grow beyond the need to write down the amount of detail needed by a trainee. I have included a checklist below to enable you to check quickly to see whether your trainee is including essential ingredients. There is obviously no 'one way' to set out a lesson plan any more than there is 'one way' to plan a lesson but hopefully the following list of points to consider will help.

Look for an overview of the teaching/learning experience:

- name of trainee teacher;
- name of class teacher;
- subject in the school curriculum;
- name of class/group to be taught;
- the day and date of the lesson;
- location of lesson in the school timetable (for example 1/3 in the week);
- seating arrangement (pairs, group, whole class);

- distribution of boys/girls and ability levels (for example 3 very able boys);
- brief background notes on the class/group (for example, mixed ability);
- SEN support needed and resources/personnel in the lesson;
- details of pupils' learning in the previous lesson;
- details of pupils' learning planned for the subsequent lesson;
- precise and detailed learning objectives for pupils in this lesson;
- details of how these relate to the National Curriculum if appropriate;
- overview of professional objectives for the trainee (for example use of the overhead projector (OHP));
- full list of resources needing preparation in advance of the lesson;
- risk assessment (for example watch for trailing lead on the OHP);
- a checklist of the standards for QTS met in the course of this lesson.

Look for a detailed account of the core activities planned. This needs to take account of:

- How will lesson content be covered and through which activities?
- What are pupils required to do during activities – for example read, write, speak, listen?
- What does the activity entail in terms of timing and teacher–pupil interaction?
- What does the activity requires in terms of resources for it to succeed?
- How will I check that pupils understand what they are meant to do?
- How will I know what they have learnt/if they have learnt?
- How does each core activity relate to the departmental/National Curriculum (areas of experience, all programmes of study and attainment target levels).

Look for an overview of the extension activities planned. This needs to demonstrate evidence of differentiated planning for:

- the more able pupils in the class;
- the less able pupils in the class;
- contingency plans if a core activity takes less time than planned.

This section may refer to ready-prepared differentiated resources – worksheets or textbook pages. The objective is to encourage trainees to expand their range of teaching strategies so they become more flexible and less constrained by clinging relentlessly to a fixed lesson plan. You may decide to introduce this section of the plan once the trainee has already experienced success in planning and teaching with A, B, C. Similarly this section may be expanded by you to incorporate detailed planning of any homework activities relating to and referred to in the lesson plan.

Questions to ask yourself:

- How effectively do I plan my own lessons?
- How can I show my trainee in my teaching how important it is to plan?
- How can I work with my trainee so we both plan more effectively?

Good practice in planning with your trainee:

- Explain the significance of good planning in creating successful lessons.
- Model good practice in lesson planning and discuss your planning process.
- Point out that even you cannot and do not plan for every moment of the lesson.
- Demonstrate the need for every plan to be informed by sound assessment.

Chapter 8

# Stimulating reflective practice

In this chapter we shall be considering the following aspects of mentoring:

- Understanding the nature of reflective practice.
- Engaging in reflective practice with your trainee.
- Questions in relation to reflective practice.
- Identifying good practice in stimulating reflection.

## Understanding the nature of reflective practice

'Reflection' is a term rather like 'family' or 'teacher'. Everyone happily bandies it about as if it had a universal meaning. In fact it is in such common usage that many researchers as well as trainees and mentors seem to assume that it means the same to everyone. They use it rather like a kite mark on electrical goods. If it has 'reflective practice' written on it, it is bound to be good. The fact is that it may not be good. For some people 'reflection' signifies just 'thinking about' something; for others it carries the implicit message of 'analysing what has occurred' and for yet others it heralds a combination of both a creative tension and a decision to improve practice. It is in this third sense that we will look at how to manage reflection.

A mirror reflects – it can show us what we look like – as others see us. Correct? No. We see what we look like in relation to the quality of the mirror surface. Where there are flaws our image is distorted. The reflection is only as the mirror surface will allow. Even from a good-quality mirror, the image is reversed. If we transfer this metaphor to mentoring it leads us to understand that where a mentor reflects back a novice's teaching in a lesson this reflection of practice can only be as faithful as the sensitivity and perception of the mentor will allow. Similarly if we are using ourselves as the mirror to examine our own actions, the quality of our reflections is again determined by our ability to reflect. Flaws and prejudices distort the image.

Just because a mentor and a trainee work together with the intention of reflecting there is no guarantee that reflection is actually occurring or that it is leading to improved practice. Thinking is a skill that needs to be practised and refined. Reflection involves selection in thinking and is dependent upon the quality of the reflector's capacity to communicate and to use thought to determine an improvement in action.

## Is all reflection productive?

The simple answer is 'no', or at least 'no, not necessarily'. For reflection in mentoring to be productive, it must produce a change in a teacher's attitude and in classroom practice.

Reflection in the context of initial teacher training has become a rather fraught issue. This is partly because you hear 'reflection' used as if it was static, a neat manicured process to apply throughout initial training. In fact it is messy, and untidy, it is unpredictable, it will not always come because, like a muse, it can be illusive. You can assist in its growth in a number of ways (Mitchell and Weber, 1999) but it cannot be forced and it cannot be acquired on demand.

## Engaging in reflective practice with your trainee

At this point it is useful to think about stages in trainee teachers' professional development because the nature of the stage will largely determine the nature of the reflection. If the aim of reflection is improved practice it is important for both to appreciate the conundrum that mentoring changes over time. As we consider the stages or aspects of a trainee teachers' development in depth we shall refer to broad categories of reflection. There are generic and particular kinds of reflection appropriate to different stages suggested by Furlong and Maynard (1995).

Just as there is no single definition of reflection so there is no single focus for reflection in mentoring. At their initial interview, potential teachers are selected on the basis of their existing subject knowledge, their personality, their experience of working with children and their pedagogical content knowledge

(Shulman, 1986). This is the capacity to teach what they know as subject knowledge and their ability perceived by the interviewer to reflect and use reflection for improving what they do.

Once they begin training, your ability to inspire them, to reflect and to use that reflection to improve teaching, should be at the forefront of your concerns. Reflection will be focused differently according to the stage of development of your trainees but if we use the stages identified by Furlong and Maynard for developing teachers as a starting point, we can begin to identify foci for developing reflection

### *Stage one: early idealism*

Since the distinguishing factor of this stage is idealism, appropriate reflection should address this, in part, as a potential problem but largely as a potential strength.

As you guide and stimulate your trainees' curiosity about teaching you need to retain their sense of excitement, of wonder at engaging in a new profession, while gently encouraging them to harness their thoughts to systematic exploration of the new environment that is your school. Often, in initial teacher training, trainees are set challenges at this stage to think about teachers who have particularly inspired them or teaching strategies that they have used. This is good, to a point, but where does it lead mentees on to develop? Is the challenge suggesting that imitating the role model will bring success? How can novices understand what constitutes good practice, how can they come to understand that 'knowing that' is not 'knowing how'.

Herein lies your real challenge as a mentor. This is a time for a trainee's exploration. The role as model for your trainees must be set alongside other role models for 'good practice' if they are to appreciate that there is no one way of 'good' teaching. Reflection should be leading to the realization that teaching is an evolving interaction that can be successfully managed in different ways. Your trainees need to come to an understanding that teaching is not as easy as it sometimes looks; that experienced teachers are not 'good' all the time; and to understand why they are successful at all.

This is also a stage for reassuring mentees that they have the potential to become good teachers, for setting in a store of self-esteem upon which to draw in the turbulent next stage. Reassurance might take the form of drawing out and identifying those life skills that they already exhibit, or exhibit if the selection process has worked:

- good study skills;
- good subject knowledge;
- good social and management skills.

In short your challenge is to recognize and work with your trainees to recognize that they come with skills *but* that these alone cannot bring about 'good' teaching.

The best way to start perhaps is to undertake an audit of the three 'goods' outlined above and to encourage your trainees to consider how to transfer them to teaching.

### *Stage two: survival*

Trainees usually hit this stage head on once they start to teach. Just as an expert teacher will prepare pupils to rise through attainment target levels in the National Curriculum so you should prepare for a trainee to pass through this stage as unscathed as possible, retaining a passion and a self-esteem borne of achievement. When I started teaching, it was a given that there would be a 'baptism of fire' where I would survive or expire by drawing on my own resources. In fact I was incredibly fortunate. I had a wonderful mentor called Bill Sugarman before mentors were in vogue. He reassured, challenged and held my hand as I got to grips with survival! I had faith in his judgement and his trust was reciprocated. We reflected – together!

Sometimes mentoring can begin to go wrong here. If trainees encounter problems and feel intimidated that 'everyone knows more than they do', they can regress to a sense of failure. Properly prepared, trainees will already have coping skills to enable them to pull through. These might include time and stress management but, if not, training could usefully be provided as an investment against regression by the trainee. From the audit in stage one the mentor needs to bring about reflection in trainees that will enhance their growing ability to flourish.

The craft of the mentor here is to enable mentees to call on those coping skills they have developed from the time before they embarked in teaching. Quite frankly, if they do not have coping skills that they can bring to teaching this is an unsuitable basis for entering the profession. Sometimes hard decisions have to be made at this stage and if trainees are unable to reflect and draw on their previous personal and professional resources for coping they are not yet ready or are ill-suited to teaching.

The trainee needs to reflect on the exact nature of problems but significantly needs to do far more than this. He or she needs to reflect on strengths as well as difficulties – 'what am I good at', in addition to 'what am I not so good at?' (and 'what I am going to improve upon?') Simultaneously the mentor needs to be reflecting on the same issues but planning ahead for the next stage, preparing the way by creating opportunities for the trainee to appreciate and develop strengths and to face difficulties in security.

### *Stage three: dealing with problems*

The focus for reflection at this stage largely depends upon the nature of the challenges facing the trainee or the nature of the challenges being set for the trainee by the mentor. The objective at this stage is for the mentor and trainee together to face those problems threatening the trainee's survival and to create not only targets but action plans for improvement. The trainee will be teaching now and the temptation is to seek out quick fixes – but again knowing that is less useful than knowing how to do it. Throughout the mentoring process there needs to a reinforcement of what the trainee has already achieved, what the trainee is already capable of achieving and diagnostic assessment of the trainee's work as a basis for planning ahead. The trainee must move towards and must be moved towards autonomy. Dealing with problems is not achieved by reliance on the mentor to sort things out – but on enabling the trainee to see how he or she can solve problems without recourse to dependence on the mentor or other teachers. Dealing with problems incurs taking responsibility and you as mentor can initiate reflection by asking your mentee 'How did this work and *why*?'

### *Stage four: plateauing*

Your trainee has learnt the basics and has sufficient pedagogical knowledge to deal with problems and to feel quite secure. The challenge facing both partners in the mentoring relationship is how to leave the secure ruts of the plateau. The good thing about a plateau is that it is high and flat enough to survey the surrounding terrain. There is a sense of achievement – in scaling the sides of the plateau – but a plateau can be featureless and windswept. You have to stimulate your trainee to move on . . .

### *Stage five: moving on*

Now the craft of the mentor is to reflect on stimulating the mentee to leave the security of the plateau, to cede the relative ease of life there and to look for challenge. It is not for you alone to provide the opportunities now; your trainee should seek out challenges: new heights to scale. The balance of reflection is shifting. In the early stages, you initiated the challenge to move your trainees on. Now your trainees are initiating the reflection. While you are engaged in reflection as a basis for summative assessment your mentees are engaging in formative assessment of their own practice against their potential; weighing up where they are and where they go on from here.

## Questions in relation to reflective practice

### *Why should I reflect?*

By reflecting on what we do, we are opening up our practice to question with the specific intention of improving it. Reflection is an active and formative process. Through reflecting we begin to seek answers in our work to questions like these:

- What do I do well? (This enables us to learn to value our own practice.)
- How can I do it even better?
- What don't I do very well? (This enables us to learn to accept our own fallibility.)
- Why do I need to do it better?
- What action am I going to take so I do it better?
- When am I going to start?
- Can I do it on my own or can I get help?
- How will I know I am doing it better?

Reflection is the stuff of learning. Through reflection we can understand how what we do impacts on us and on others. Without reflection we are left with vague impressions, half digested truths and dependency. By learning how and why we do things we are in a better-informed position to decide how to bring about improvement. Reflection is an over-used term that is sometimes misunderstood.

Just thinking is not enough to bring improvement; it has to engage with real action!

### *What should I reflect about . . .?*

Whitehead (2000) challenges us to adopt a stance of challenge towards ourselves by his focus on questions like 'how do I improve my practice?' and 'is there any evidence that I have influenced your learning for good?'

To Whitehead's challenges I would add others drawn from the research that I and others have carried out into using visualization as a means of understanding and improving teaching:

- How do I perceive myself as a teacher?
- How am I perceived by others as a teacher?
- How do I want to be perceived as a teacher?
- How do I actualize my vision of myself as teacher?

### *When should reflection take place?*

Reflection should be encouraged at each of the stages outlined above – it is after all part of the professional development of the trainee and so the observation that trainee teachers cannot be expected to reflect until they have sufficient experience upon to reflect is misleading. Reflection can and should occur throughout the initial education of teachers for without reflection, there is no education. But how can this be? Education in its classical form means 'rearing' (from the Latin *educatus*). I am attracted by the vision of 'rearing' trainee teachers! It has connotations of growing up and growing on. Reflection is the nourishment of the educative process. *Reflection is the basis for learning: it questions the automatic response and thereby enables transference of skills between activities undertaken.*

### *So what if we relate this theory to a practical example?*

If you want to educate the trainee in how to plan lessons just saying 'Do this' may provide a quick fix but it does not develop a skill for planning. Your trainee needs to know not only what to do but why this strategy is likely to be effective for this class. Otherwise planning becomes mechanistic and detached from interaction, and that exquisite sensitivity that good teachers display in planning and teaching their pupils becomes just a performance played out in front of a detached group of children.

For your trainee's professional development to get under way systematically, it is vital for you to select – and to enable your trainee to select – those aspects of practice that can form the basis of improvement. If you feed back every detail about the practice, your trainee will soon feel overwhelmed. The craft of mentoring is learning the creative use of appropriate selection and not random feeding back for reflection.

### *Who should be reflecting?*

It should be clear from the observations and suggestions relating to reflection under the stages sketched out above that reflection is an activity for you, the mentor, as well as for your trainee. While you are engaged in reflecting upon your own and your trainee's practice, your trainee should simultaneously reflect on his or her own practice and on yours as a teacher. Thus, the mentor should be seeking to improve his or her teaching and to influence the mentee's practice in a positive and formative manner. Simultaneously mentors should be using every opportunity afforded by mentoring to improve their work in school, and this includes learning from trainees.

### *How can reflection be stimulated?*

Reflection requires a place of calm, a nest of security, and a pool of knowledge, and yet ironically it grows from tension, from contradiction and from striving to balance.

Without a certain tension there can be no motivation to improve and to develop. But how might you demonstrate that you are engaged in reflection as a reflective role model in the hurly-burly that is the norm for teaching? First of all you need to *be* a reflective practitioner before undertaking mentoring just as much as your trainee. There needs to be a process of reflecting back so there can be an audit of skills and areas for professional development. There are many ways of stimulating your trainee to reflect, but you may find the following a useful starting point after a lesson:

- Can you explain why you did X at that point in the lesson?
- Why do you think X occurred?
- Why didn't X do as you said, do you think?
- How could you have made sure everyone understood what to do?
- Where was the opportunity for the class to learn about X in your planning?

---

## Illustration

The mentee observes a lesson taught by the mentor. Only the events of the lesson are recorded – in writing, or on video or audio recording. As the trainee and the mentor reflect after the lesson the mentor demonstrates or models her practice of reflection:

> When I saw X happen at X time, I thought this indicated . . . I realized when X happened later that my earlier reflection was correct/incorrect . . . What did I think when Y happened?

And she invites her trainee to join her in that process:

> Why do you think I did Z? This is what I had in my mind at that time. What could I have done so that W occurred, do you think?

The mentor is constructing a 'virtual' lesson in the light of the observation and initiating the mentee into sharing the insights. You cannot expect your trainee to be reflective without demonstrating your own ability to reflect, even if you are a novice at it too, and opening up your practice to scrutiny. You need to examine publicly your own strengths and shortcomings in your teaching to demonstrate to your trainee that knowing one's self is the basis for insightful teaching and it is 'safe' to reveal that you sometimes do not know. The process of co-enquiry is the basis of good mentoring.

---

### *Where should reflection lead?*

Reflection should lead to action, if we use the term in the sense of the opening paragraph of this chapter – but how?

Reflection with a view to improvement means avoiding wallowing in despair – depression is a luxury neither mentor nor mentee can afford. It means knowing failure alongside knowing previous success, and knowing that success is attainable in future. In terms of learning it means scanning through your memories for similar situations from which to draw inspiration, situations where you not only coped but flourished. Where there is no suitable parallel it means imaging oneself in a situation where the difficulty is being posed and analysing what, in this idealized state, is within your grasp to improve within the actual lesson.

It may be as simple as remembering to smile, or to adopt more assertive language. It may necessitate learning new skills in preparation for taking a lesson. The reflection you and your trainee undertake is to identify what exactly needs to be done before, during and after the teaching session, and is the preparation for improvement.

---

#### Illustration

B has been a mentor for several years now. She has worked with strong mentees who have been appointed to teaching posts in her department. In fact it is not so much that she takes on strong mentees but that she has the ability, through her own reflection, to understand how what *she* does as a teacher effects what her *pupils* do as learners – and she learns from them. In order to do this she has video-recorded many of her own lessons and has analysed thoroughly and methodically how it is that her teaching has or has not brought about learning in a particular context.

#### Commentary

Using her own experience of reflection, she is an excellent role model for her trainees and she has the insight to analyse their practice. Her particular skill is in recognizing the importance of context and pointing this out to her mentees. Not all hear what she says at first but they will all learn in time. She stresses that a particular strategy is only successful in so far as it meets the needs of a particular class in a particular lesson. She draws parallels for the mentees between situations to enable them to consider a range of possible teaching strategies to elicit learning from her pupils – she seeks the novices' feedback on their own teaching as well as her own as a basis for their improvement specifically but, incidentally, for her own.

---

So what *is* reflection?

It is thinking about how what you have done, are doing and will do makes you a more effective teacher.

*So can every reasonably intelligent person reflect?*

Yes, but some need more training and practice than others.

*So can reflection be learnt?*

To a certain extent. The strategies for stimulating thinking can be learnt but for this thinking to translate into improving teaching there needs to be action.

*So how can I translate thinking into action?*

Undertake action research into your own teaching.

*So how will undertaking action research into my teaching help my trainee to teach?*

By demonstrating your readiness to learn and your capacity to educate yourself you are modelling your good practice for your trainee teacher.

Questions to ask yourself:

- Do I use reflection to improve my own practice?
- What evidence do I have that it does lead to improvement?
- How can I model effective reflective practice for my trainee?
- How can I stimulate my trainee to reflect and improve our practice?

Good practice in stimulating reflection:

- Be a reflective practitioner yourself.
- Be prepared to help your trainee to learn to reflect.
- Be ready to reflect openly and honestly on your own areas of improvement.
- Be improving your own practice as well as helping trainees to improve theirs.

# Part 3. MENTORING IN A SCHOOL CONTEXT

Chapter 9

# Making the most of partnership

In this chapter we shall be looking at the following aspects of mentoring:

- Contexts and possibilities for initial teacher training (ITT) partnership arrangements.
- Roles and responsibilities in partnership arrangements.
- Engaging in creative partnership with higher education institutions (HEIs).
- identifying good practice in ITT partnership.

## Contexts and possibilities for ITT partnership arrangements

When Sheila Lawlor attacked HEIs in 1995 for teaching novice teachers 'trendy' theory, she was voicing a dissatisfaction that hit at the very heart of arrangements for ITE that had existed for many years. There had been some exceptions to the largely HE-based model in which schools provided opportunities for practice but theory resided with HE. The Oxford Internship model and the Sussex University partnership were the models for post Circular 9/92 partnership arrangements when schools became major stakeholders in ITT provision. When Circulars 9/92 and 14/93 came into effect, there were strident cries from some HEI lecturers that schools would not be able to train teachers as effectively as they had. ITE would be a bolt-on extra and the senior teachers envisioned as the harbingers of good ITT provision by Lawlor would be found lacking. For others, change was an opportunity for better practice. At last, both partners would be working in unison to maximize their strengths.

In the early days there was a pioneering spirit in many schools that was welcomed by HEIs. In others, there were accusations of training on the cheap, of 'sitting with Nellie' and a feeling among teaching staff that this was 'not what we're here for'. Most schools still looked to HE to take the lead by telling them what to do in mentoring, little realizing apparently that they now had a much stronger gatekeeper role and an input into the profession of teaching never before residing in their quarter. Since the move to greater school involvement in ITT, the developmental potential for both is increasingly being appreciated but there is still a considerable way to go.

Good communication lies at the heart of effective partnership. Roles need to be clarified and boundaries established before partnership is launched. There needs to be a clear understanding about how to deal with problems and this understanding needs review on a regular basis by both partners. Clarity relating to role expectations is essential to good practice for all. In particular, schools and HE providers need to negotiate their respective roles within the assessment of student teachers. If there is disagreement over a grading, who has the final say? If there is a problem with a novice teacher's progress, who is responsible for sorting it out?

The most effective partners are those who have a clear understanding of the role that they and their partner have in working together. This can only grow through the process of working together. Detailed documentation is useful as a checklist and a reference point but partnership is about people rather than pages; respect as well as reference. School-based mentors should be fully aware before they undertake involvement that the onus for contacting the tutor rests with both parties. Too often both have to badger for contact to be maintained. A simple but effective system involves the mentor ringing, faxing or e-mailing the tutor every fortnight during the block teaching practice unless there has been reciprocal contact during that fortnight. A written message is more use than a spoken one as it leaves a text report for both sides to refer to. A message left on an answerphone is better than nothing.

In partnerships there is normally a course committee or policy committee where representatives of the school meet the HE staff to discuss issues that arise in the process of working together. Do all mentors know who is their representative and how to raise issues that concern the course as a whole? They should. Do all make use of its communication network? Mentoring can be an isolating experience and so it is important for ITT partners to accommodate mentors' needs for support in their work. Mentors need mentors and partnership has to make room for this need.

This opportunity for mentors may be afforded during mentor liaison meetings or mentor development sessions, typically once a term, when all the mentors in a particular ITT partnership talk over their work and plan for the next stage in the year.

This is a vital time for colleagues to meet up and share good practice as well as to use the expertise of university tutors to enhance their subject and ICT skills. How frustrating, then, when mentors are invited to meetings but fail to take part. How can partnership be effective unless partners work closely together?

When a mentor misses a meeting he or she should ensure that a colleague from the same department goes instead. If not from their department then at least from their school, so valuable communication is not missed. At present, attendance can be as low as 35 per cent even when partnership meetings are advertised well in advance. This is not professional practice. Similarly, schools can be left in limbo when tutors fail to show up on the day and at the time agreed to review a novice teacher's progress and provide support for the mentor. Again a surrogate should be found. Where attendance at meetings is not possible an alternative strategy should be suggested. The responsibility for arranging this lies with the partner who cannot be present.

In order to ensure a shared understanding of what ITT provision involves and to see how this sharing impacts on mentoring it is vital to consider the context of mentoring as well as the roles and responsibilities of partners. How might these be defined?

## Defining respective roles of HEIs and schools in ITT partnership

Shared school and HE responsibilities:

- arranging cross-phase activities for trainees;
- providing references for job interviews;
- meeting to agree procedures and standards for assessment;
- marking assignments for subject and general professional studies;
- ongoing formative assessment of novice teachers' teaching;
- interviewing prospective trainees for the ITE course;
- designing and planning the course structure;
- moderation in the event of a failing student;
- moderation of all trainee's progress by delegated committee.

School-based responsibilities:

- providing an in-school induction programme;
- organizing suitable classes for the trainee to observe and teach;

- providing subject-specific mentoring;
- providing general education and professional mentoring;
- setting up in-class training with an experienced member of staff;
- planning, observation and debriefing of lessons;
- form tutor mentoring;
- opportunities for developing information and communications technology (ICT) skills in teaching pupils;
- regular mentoring sessions on a weekly basis;
- offering opportunities for extra-mural activities;
- offering opportunities to attend staff and parents' meetings;
- offering opportunities to assist with assemblies;
- assistance in meeting the standards for qualified teacher status;
- designing the in-school programme for the trainee teacher.

HE-based responsibilities:

- designing the in-HEI programme for the trainee;
- course administration including grants and bursaries;
- providing training for new mentors and professional tutors;
- setting up and running liaison meetings for mentors;
- providing secretarial support for course administration;
- moderating school-based assessments of progress;
- providing validation for the training course;
- printing and distributing course documentation;
- contact with the graduate teacher training registry (GTTR);
- recruiting schools to partnership;
- training mentors;
- recruiting trainees to teaching;
- moderating marking of assignments and research projects.

It is an inescapable fact that in all partnerships there are potential flashpoints – and in the HEI/school partnership the resurgent flashpoint is one of funding. How much money is allocated to cover novice teachers' use of photocopying resources? Are mentors to be given financial rewards or time in lieu or both? It is not the HEI's responsibility to dictate how schools use the funding that they receive. However, some mentors do feel uncomfortable at liaison meetings where some colleagues are paid for their work by schools and others are not. This matter does need consideration at school and mentors have a responsibility to air their concerns with senior staff. Higher education institution tutors are likely to be very sympathetic to the in-school needs of mentors but they cannot become embroiled in controversy about who should get rewarded for what.

## Sharing a language for partnership

In ITT the host of terms describing the basic roles and responsibilities of the main stakeholders in partnership often remains confusing and bewildering. It is ironic and unacceptable that there is such a lack of clarity within the process of teacher education. It is usually in conversation that partners realize that they have differing conceptions of 'reflective practice' 'assessment' and 'responsibility'; all fundamental issues and all requiring clarification. So often issues are slightly misunderstood and it is only when problems arise further down the line that it can be seen how slippage in communication has occurred. You could say 'well, it is up to them to say what they mean'. Higher education institution colleagues are likely to feel the same because the fact is that it is the responsibility of both parties. Close liaison leads to a shared language for ITT.

## Sharing a culture for partnership

Similarly, sharing a culture means that both parties in partnership have to agree to and uphold the values and procedures that underpin the practice. Where, for example, it is agreed in a mentor development meeting that all trainees will use a standard pro forma to set out lesson plans in school, then it is a rejection of the shared culture of the partnership when a mentor initially agrees and then uses a different pro forma. Keeping to agreed procedures saves time and effort where both are at a premium. Shared lesson plan pro formas for example can be used as a common basis for discussion in sessions with the visiting tutor; averting the need to ask a host of questions about yet another 'school' pro forma. It is worthwhile looking at the ITT course procedures and seeing where there can be streamlining by sharing common practice – and adhering to it. When trainee teachers are observed by teachers in school there is sometimes a partnership undertaking that a summary of the feedback following the lesson is to be given in writing and kept on file for future reference. Fundamentally, it is unfair to the trainee to do other than this and is a mark of the status accorded to their professional training.

## Your role in liaison

The place for you to raise concerns about partnership procedures is in contact with the HEI and specifically with the person charged with arranging ITT partnerships. It may be appropriate to share your concerns with other mentors and HEI tutors during mentor development and partnership liaison meetings.

Agreeing and sharing procedures is fundamental to effective ITT practice and to contributing to a supportive and open 'cultures of partnership'. Where you need to be pressing for exceptions it is more constructive to say 'what if we . . .?' rather than refusing. Where possible be prepared to offer a workable alternative if you cannot agree. It is important to recognize that partnership does mean compromise – by both partners. *Mentors and tutors should seek shared practice, shared understandings.*

Clearly both partners will have their own priorities. Expecting, for example, that the course requirements of three very different ITE partnership programmes can be brought together in a seamless way in the time allotted normally to just one ITE programme not only puts pressure on trainees in school – it can distort what happens in the HE sessions too. Both partners will have other priorities and responsibilities to attend to. Tutors who set unreasonable deadlines for assignments to be completed or who refuse to negotiate the timing of their school visits are disrupting not just their own subject area's participants but the fabric of partnership too. Similarly, mentors who delay returning reports on novice teachers' teaching obstruct good partnership.

Where there is clearly friction resulting from trying to align very different ITT programmes into a single provision in school, it would be better to offer to take two or three partnerships and run separate mentoring sessions to accommodate the needs of each one, rather than to force uneasy compromise by insisting on a joint session for trainee teachers. On the other hand there may be benefits in some combining of mentoring because trainees can learn from one another and mentoring can benefit from the expertise embedded in more than one ITT programme. Initial teacher training programmes are bound to differ and the best features of each are of potential benefit.

In the next illustration we shall see how one teacher effectively combined them. The aim for partners who are engaged in the day-to-day functioning of a shared ITT programme should be one of supportive and creative compliance with agreed procedures. Where a school takes on a multi-partnership approach diversity needs to be exploited to enrich the practice of training teachers in the school. This is more likely to be achieved where the school actively seeks partners whose programme work in a largely parallel way. If this is not the case the school could ask for more trainees from the one partnership so the coherence of the in-school programme is reinforced.

---

## Illustration

T is the headteacher of a school in partnership with three HEIs where the course requirements are similar and complementary. He wanted to work with just one HEI but had no local provider that offered trainees in all of the subject areas where his school could offer good mentoring practice. He chose the first HE on the basis of knowing the personnel – the course director had been a former member of staff whom he liked and respected. He chose the second on the basis that its programme was similar and covered other subject areas, and he chose the third because it ran at a different time of year and meant that mentors developed their skills in a slightly different programme. Some of the general education and professional studies sessions catered for all of the trainees from different institutions and some were discrete sessions where the programmes differed. He was a part-time subject tutor for one ITE course and so was aware of the 'other side' of partnership. He kept in regular phone contact with the director of studies for all three programmes, checking that his in-school programme met the different course demands in an integrated way.

## Commentary

T's attention to partnership arrangements is managed on a number of different and complementary levels. First he ensures that the timetabling of in-school sessions will not cause over-burdening of staff or of pupils in classes taken by trainee teachers – ensuring that pupils have time without exposure to ITE and protecting his examination groups in years 11/13. Second, he is concerned to maximize the human resources in his school. Mentors are drawn from most subject areas as T is keen to become an ITE training school on the basis of putting ITE within staff development. His school has already achieved the status of being awarded a Charter Mark as a public centre organization in the county and is expecting shortly to become an 'Investors in People' designated school site as well. Thirdly, T demonstrates that partnership means giving and taking. He contributes to one course as a lecturer and keeps in regular personal contact with all three course leaders, negotiating the best input for all. He is also ready to come out of partnership should any of the three courses make demands on his staff and pupils that would endanger their wellbeing.

Rather than opting out of agreed procedures 'because they don't quite fit with the programme from the HEI', try to negotiate more congruence during whole-course policy meetings by raising issues for the agenda well in advance. Give colleagues in your partnership time and space to consider ways of aligning ITT practice. Where, for example, a difference in practice between ITT courses arises that cannot be accommodated easily, send in a brief paper in advance of the whole-course policy meeting raising that particular issue so it can be circulated and then discussed in depth by partners during the meeting. Nobody has a monopoly of good sense or good practice. We need to learn from one another's programmes to build a stronger future for practice. Above all avoid just keeping quiet and 'doing differently' in your ITT practice or, worse still, simply leaving one ITT partnership without airing your views and providing an opportunity for changes to be made to enhance partnership practice.

## How can you become a better partner?

- Define your own role as a mentor in a particular ITE course.
- Define the role of your HE partners – in your school and in the HE.
- Make the most of mentor development and liaison opportunities.
- Build on continuity in the ITE programme between school and HEI.

One of the most important areas where you as a school-based mentor can make a strong contribution to partnership is by ensuring you are familiar with the course programme documentation before the novice teacher's arrival. Nothing is more dispiriting for the trainee than hearing a mentor say 'I'm not sure what I am supposed to do – you had better ask your tutor!'

Such occurrences are happily rare but they do occur, and undermine partnership. Similarly, the university-based tutor who insists on telling the mentor what to do and how to think, is exceeding the HEI's role. Much has improved since September 1993 with the advent of new partnership arrangements but there are still tutors and mentors who cling resolutely to their pre-Circular 9/92 roles and are apparently unaware of the major changes in schools' responsibilities for assessing and supporting trainees to meet the statutory requirements of the standards for QTS. If you recognize yourself among this number you have an added impetus for studying chapters 17 to 20 in this handbook, where you will find each area of the standards and its implications for mentoring covered in depth. While all ITT partnerships are accountable to the standards there can be a very considerable variation in how the partnership and the ITT programme is arranged, as the previous illustration bears out.

In the next illustration, we can see how one school responded to the call to participate in more than one ITT partnership. The crucial factors in the success of this response were the strong team ethos of the subject department, the number of staff already experienced in ITT mentoring and the size of school, which allowed for several trainees to meet different classes on a regular basis within the subject area.

## Illustration

G has been working in partnership with one ITT provider for a number of years and had offered two PGCE placements across several of the subject areas in his school. Recently he was approached by another ITT provider asking for school-based placements in a subject area where he already had mentees. His is a large school and he already has three experienced subject mentors in this one department. Rather than undertake a duality of partnership arrangements and having to double up on mentoring, he contacted the original HE provider and explained the situation. He offered to take a further trainee in this subject area on the basis that mentor training would be offered to a third subject teacher. The call from the second provider was politely declined.

## Commentary

G was under considerable pressure to take a student from a neighbouring institution. He had resources to do this but his loyalty to his current partner led him to negotiate a special arrangement for more trainees. By doing this he reinforced his school's standing in the partnership, avoiding a mix of programmes. The mentees were delighted at having three mentors with whom to work – they rotated mentoring over the year and they were also happy to have peer support available from one another. General Professional Studies sessions run by the school were concerned with just one ITE programme so no compromises were necessary. One of the novice teachers encountered difficulties but, with so much support on tap from the other trainees and mentors, soon pulled through with extra help in a demanding personal support programme tailored to her needs. All three mentors gave a little additional input so no one teacher was put under additional stress, as often occurs when problems arise.

By carefully weighing up the potential for expanding their mentoring capacity, this school was enabled to offer a more extensive programme of ITE involvement.

The pressure on finding good-quality placements for trainees continues to grow and there is a danger that partnerships will be overstretched if schools offer single placements. Welcome as they are, there are a number of disadvantages. Not least the trainees lose the potential for peer mentoring (Hawkey, 1995) and HE partners may have to pull out of the partnership rather than invest in mentor development and visiting by tutors for individual trainees. This pulling out would be to the detriment of school-based mentors who identify substantial benefits from working in partnership with HEIs.

The report published by the Association of Teachers and Lecturers (ATL) (Barker *et al*, 1996) listed a number of benefits and costs accorded by schools to their involvement in initial teacher training. The list of benefits was substantial.

The major benefits were intangible and were expressed in terms of enhanced professional development, including curriculum innovation and new teaching methods, increased job satisfaction, intellectual challenge, and stimulus to reflection. The main tangible benefits included support with extracurricular activities and the addition of new resources.

ITE was seen as an important vehicle for professional development and schools enjoyed the spin-off of recruiting 'good' students as NQTs.

While acknowledging similar benefits, a later report prepared for the Standing Committee of Partnership Administrators (Fletcher, 1999) identified crucial issues: i) we like partnership but do not always agree with our partners, ii) we like partnership but novice teachers can be reluctant to join in.

Good practice in partnership revolves around dialogue and a recognition that partners have different and complementary roles. As a mentor you may have a very different perspective from the tutor who comes to visit your trainee. You may disagree with the tutor's impression of the trainee's progress but any disagreement needs to be supported by a fair and comprehensive review of available evidence. The same, of course, applies to the standpoint adopted by the HEI tutor who disagrees with you.

Where disagreements arise it is vital to relate these to the framework of standards for QTS which are the backbone of assessment and are the best tool we presently have.

Novice teachers are reluctant to join in? What are they being asked to do? Is it reasonable and does it correlate with the demands of their ITT programme in HEI? There are opportunities for creative partnership with HEI colleagues beyond the classroom that can cement collaboration within the ITT partnership you hold.

---

## Engaging in creative partnership

In this illustration we consider examples of the creativity that ITT partnership can engender through imaginative ways of deepening trainees' teaching experience.

## Illustration

Trainees go out to primary school as a group and run activities with the pupils based on resources that they have developed within an assignment.

Trainees mentor GCSE candidates by e-mail from their university base, offering personal support through reassurance and subject guidance.

Trainees link by video-conferencing to A level pupils from the university to distant schools to increase their understanding of the GCSE–A level gap and to offer positive advice to students to support and extend their learning.

Trainees as a group organize curricular events in schools. These might include modern language theme days and assisting with curricular events like field trips and maths paper trails.

Trainees make learning resources for pupils so they can be more autonomous in mastering aspects of the curriculum that they find difficult to grasp in class.

Trainees can carry out research in collaboration with school to inform school policies and the school development plan. This might focus on use of ICT.

Trainees often come to teaching from other careers and can sometimes use their contacts beyond the school to organize events for the pupils in school. This might include links with journalism, with service industry and business.

Good partnership extends beyond the teaching block when novices are working in school. As a mentor it is important for you to be proactive in partnership by suggesting ways of creatively collaborating with HE colleagues. Your time may be limited but so is theirs, yet together you can maximize the potential for partnership.

Questions to ask yourself:

- What is my role and what are my responsibilities in ITE partnership?
- How can I improve my contribution to partnership for ITT?

Good practice in ITE partnership:

- Build a complementary role with your HEI tutor.
- Engage openly and fully in partnership arrangements.
- Bridge differences in language and culture by dialogue.
- Make opportunities for creative collaboration for your trainee.

Chapter 10

# Mentoring within a SCITT consortium

In this chapter we shall be looking at the following aspects of mentoring:

- What is a SCITT?
- Roles and responsibilities in a SCITT.
- Delivering a programme for ITT.
- Identifying good practice in a SCITT arrangement.

## What is a SCITT?

Currently there are 52 consortia in England training approximately 2 per cent of new teachers (*Times Educational Supplement,* 3 March 2000). School consortia that work together for SCITT (school-centred initial teacher training) are already well established, running successfully in some regions and embodying good practice. Often these consortia are set up where there is no other ITT provider or where existing providers do not offer the subjects that schools require. Some SCITTs, as they are known, exist in partially devolved partnerships with higher education institutions that offer validation and quality control. SCITTs are not necessarily in competition with HEI/school partnerships.

It is important to realize that there are usually strong similarities between SCITT and the HEI/school partnership routes into teaching we considered in chapter 9. In both SCITT and HE/school partnership, trainee teachers may be settling into their school from the first week of their course. In both there is likely to be a well-planned and well-structured training course that conforms to rigorous standards set by OFSTED and validation boards. In both, school-based mentors work closely with trainees in allotted protected time and give training above and beyond this as the need and opportunity arise. In both routes into teaching, all the trainee teachers have to demonstrate competence in relation to statements in the teacher training agency's (TTA's) standards for QTS.

## Roles and responsibilities in a SCITT

A well-run SCITT can increase the professionalism of the entire school population as well as providing excellent training for novice teachers. Becoming part of a SCITT consortium raises a number of issues relating to the culture and purpose of a school:

- Where does ITT fit with the responsibility of educating pupils?
- Is a school appropriately staffed to undertake the role of leading a SCITT?
- Has the SCITT sufficient specialist knowledge to administer ITT courses?

The success of the venture depends upon recruiting, training and supporting staff who are highly committed to it. It also requires a certain vision and strong determination.

These qualities are strikingly present in the mission statement of this SCITT:

> Above all we believe that involvement in training will enhance the professionalism of schools and lead to an improvement in the overall quality of teaching and learning nationally.

But how does a mission statement move from rhetoric to reality? In the following account we gain an insight into this process of transformation. Becoming a SCITT can literally transform an organization over a period of time and in a number of ways; this is a brief account of my own experience as I moved from being a mentor to being a researcher; and I witnessed at first hand the transformation of just such a school organization.

---

### Illustration

When I undertook to become a mentor I was less than attracted to the location for my training. A tough inner-city school, a rather dilapidated building, an area of social disadvantage and high unemployment . . . but I owe my enthusiasm for mentoring and much of my understanding about what it entails to the staff at that school. Some five years after my training as a mentor I was privileged to return to the same school to undertake research into the running of a SCITT. It is this research that forms the basis of my pool of good practice here.

The OFSTED report for this lead school in 1997 reads:

> X is a comprehensive school for pupils aged 11 to 16. It provides education of good quality in a caring and supportive environment and enables pupils to progress well, both in their studies and in their personal development. The school serves an area to the west of X town centre, parts of which have much higher than average levels of disadvantage. More than 90 per cent of pupils are from ethnic minorities and the vast majority speak English as an additional language . . . The percentage of pupils achieving five or more higher grades at GCSE has been at or above the national average for boys and above the local average for boys for the last three years. As the overall attainment of pupils on entering the school is well below average these figures indicate that the school clearly adds educational value during the time the pupils spend there and that pupils tend to do better than might be predicted from their earlier achievements. The school is successful for bringing able pupils to high levels of achievement across a range of GCSE examinations.

## Commentary

This glowing OFSTED report highlights how the school's role as a lead school in a SCITT has substantially enhanced its effectiveness as a provider of education for pupils. Far from detracting from this function, a commitment to ITT has constructively contributed through its caring and supportive environment'.

So if commitment to a SCITT is potentially such a transformational experience, why have SCITTs been given a relatively poor press? Some have not adjusted to deliver a high-quality programme for ITT without HE support and some roles and responsibilities remain unclear. For others there is insufficient staffing to support the task of administering a SCITT consortium. The answer is not straightforward.

Many of the criticisms levelled at SCITTs can also be laid at the door of partnership. A paper prepared for the Open University validation services (De Vries, 1996) underlines this paradox; comments in italics are a commentary on the extracts of original text quoted.

Teachers are in a role conflict situation having to report to the SCITT consortium co-ordinator but also up the line of their school hierarchy.

*But the same can be said of mentors involved in HE partnerships.*

Senior staff could suffer burnout or lose interest.

*True – and many do in all aspects of ITT and continuing professional development (CPD) but this is less likely where there is appropriate vision, delegation and the success is evident.*

Mentoring was seen to be inadequate to address the problems of mentors' insularity –

*But in good consortia there are regular full-staff meetings in addition to subject groups every month – ample time to share and debate.*

This report is very useful as it raises a number of other issues for SCITTs to consider:

- Getting the right staff for the job – too often success depends on one person's vision and hard work rather than enabling a whole school.
- Dealing with novice teachers in difficulty – are there sufficient appropriate ways of dealing with the problems that ensue?
- The perennial problem of shortage of time – are staff able to do ITT?
- The financial headache of running a costly operation and paying staff.
- Issues relating to quality assurance – selecting schools and mentors.
- The content of the course – does it prepare staff to teach only in this location, in this consortium of schools?

---

## Delivering a programme for ITT

School-based mentors and the staff who provide the education studies programme are all involved in aspects of assessing trainees' progress. There is a progressive introduction both to the theoretical ideas that underpin professional practice and to the various dimensions of the professional debate about teaching and learning. To ensure that everyone understands the process of initial teacher education, and this scheme in particular, there is a shared staff and trainee handbook for all to use.

A feeling of shared purpose is essential to operating a successful SCITT – and a clear indication of who does what and how this relates to the whole. If we look at extracts from interviews with ITT providers in a successful lead

SCITT school we can gain some insight into the culture of professionalism they have crafted in a SCITT.

## *An interview with J*

From the outset the scheme wanted an LEA representative . . . they asked the LEA to second me for two days per week. I do all the financial side and the quality assurance. B does all the practicalities . . . (and monitors students' feedback) through a representative from each subject group, with a monthly meeting with them, but also questionnaires. The trainees are affiliated to the university and we use their resources centre. We pay money to them for the use of the centre but we tell them what to spend it on. SP coordinates the programme and delivers a fair amount of it but we bring some outside people in. On some Monday mornings I do the second half of the morning with the trainees. I've got a perspective of schools across the authority. We've done a review of the strands of our course and there's a sheet where they're laid out. In the first term those are the things a mentor is looking at so we have built them in. What's mattering in mentoring all the time is going at the right rate for them – it's the mentor who is in the key position to pull that together. Very early in the classroom they're in a team teaching situation and as both they and their mentor feel comfortable they'll start to do things – having been party to the planning. Towards the end of their first term they'd be taking responsibility for teaching an entire lesson – the handbook lays this out clearly and there are clear guidelines for everyone.

Each subject co-ordinator with their team developed their own course (for subject application and assessment) but when we did this review we saw the need to have certain common strands at certain times. We asked ourselves: 'What coherence is there within each strand? What is there in terms of continuity right across the course?' At the quality steering group we looked back at our targets for development to see how far we'd come, how we were going to maintain those that were priorities for us. The things that can make a department unsuitable for joining a SCITT are . . . where you haven't got the right attitude by all the members of your department towards supporting someone. Whatever OFSTED reports of schools there were I went through carefully and I sent each subject co-ordinator all the strengths from the departments which were already on board . . . it's the subject co-ordinators who determine the placements. How do you put ITE at the centre of the school's operation? It's to do with the impact it has on the individual . . . you can't just have certain subjects. You've got to allow a secondary SCITT group to work in every subject area – otherwise you're dividing your school.

## *An interview with B*

The letters of rejection (to unsuccessful candidates) give three reasons why they failed – we try to make it as helpful as possible. The person who carries out the screening interview (for applicants to the SCITT PGCE) knows all the mentors personally and if they don't know them the co-ordinator will spend over an hour with that mentor getting to know them. The matching (of mentor to mentee) is definitely by putting people with people that the co-ordinator thinks they'd get on well with. Mentors come and join in the screening interview and at that point some mentors say 'I particularly like X so if you'd like to put them with me that's fine.'

Following this interview the person has a placement interview and then following that interview both the mentor and the trainee are allowed to say 'no'. The screening interview takes a minimum of half an hour.

## *An interview with M*

I'm the Broad Curriculum Studies tutor and I teach most of the sessions, the twilight (after school) session on Thursday evening. I'm also a mentor for science. I sit on various committees like the Quality Steering Group and the Subject Co-ordinators' Group. Mentoring is an hour a week – that's the official time but there's probably another hour in 5 and 10 minutes scattered through the week, every time your trainee does a lesson. The second marking is an enormous job – every single piece of work. All students do the same four assignment – they're more tasks, not essays. [What are the main benefits to the school?] To the school it's increasing the professionalism of the profession. The crucial thing is that a profession regulates itself, sets its own training and standards, both academically and morally. The children are responding to the increased professionalism. The main benefit to the school is having another adult in the room all the time to start with. It's wonderful to split the burden of assisting the children when they get stuck.

Teacher morale has been enormously improved. We live in times when teachers are constantly being told they're not doing a good job – and that has a corrosive effect on morale. You're always feeling 'I'm falling short' but through this (SCITT) scheme, and looking through it at length with other equally experienced colleagues, we are starting to build up our own set of knowledge [about] what is possible. The SCITT scheme is grounded in total reality and we can never forget that. With our students we can't bluff and if I'm telling my trainee 'this is what you must do', I can't get away with that unless I'm doing it. I can't give unrealistic expectations.

### *An interview with B (SCITT co-ordinator)*

We've got the language and trainees are in a better position than a lot of people because they've been immersed in it. It's not only a common language in school – it's been adopted by the LEA throughout the county and we take responsibility for updating it. The fact that this language has seeped in has given us the possibility to talk about professional issues without saying 'what do you mean by that?' But also it's opened up the institution to the fact that new concepts are flying in and out all the time – so you could go down to the staffroom right now and talk about Gardner's multiple intelligences and people would sit and listen and have something interesting to say. You couldn't have done that five years ago.

The most powerful dynamic from a school professionalization point of view is overcoming what we call the Vygotskian Fracture – if the professional studies, the students and the school form a triangle, if you don't link the mentors with that professional studies element in a living way, you'll create a definite fracture. I work tremendously hard yet I don't feel tremendously stressed. We'll go into research. Part of my ambition is to sit around a table with a group of professionals where we're talking about the particular methodologies we're going to use to deliver particular forms of curriculum because we have the evidence that if you use this methodology in this particular way it leads more effectively and efficiently to these sorts of outcomes.

---

## Commentary

What strikes one immediately in the transcripts above is the sheer professionalism, the commitment and the expertise that underpins the practice in this lead school. There is meticulous planning in the preparation of the course, in clarifying roles and allocating responsibilities. Expectations of trainees and others involved in the delivery of this SCITT programme are clear and realistic but high. Significantly, there has been careful consideration of procedures to follow should any problems arise.

---

In the next set of transcript extracts we can share the perspectives of some of the mentors involved in working in the classroom within this SCITT consortium.

### *An interview with K*

What brought you into mentoring here?

B suggested it to me, but I did it really as a means of career progression When we went through the appraisal process it helped me in my teaching. For our department it's helped us to analyse what we do and we've obviously got to make sure that our procedure and professional standards are good and that we pass that on to the students who we're with. This whole ITT course was B's initiative. B has set up a good system. The over-riding factor [in teacher training] is professionalism.

### *An interview with A (mentor)*

After school my trainee would sometimes want feedback – it didn't fit the hour at all – so I'd stay for an hour after school each day to help her to plan, and then they phone you at home. I could see she would be a good teacher so it was worth the effort! We have a set of questions to ask at interview but personally I think first of all I like to see the spark of the person there – Mentoring? It's listening, talking, sharing, cajoling, being interested, it's developing the whole person and helping them to progress.

I tell all of them 'when you're teaching you never stop learning yourself'.

B has always been available – I think you need one person at the top who has the enthusiasm to pull it together, not just work but the people.

### *An interview with W*

. . . it's something deep inside you that tells you if a placement is going to be right – I think it's enthusiasm and a commitment, which has to be very big. That's what mentoring is about – someone who is there, who you can trust and really talk it through with and someone who shares where you're going and where you want to go. I think it's a lovely system! We have staff meetings and B fills us all in, then we have subject meetings and before we do anything else B has us go round and tell the group about our trainee – if a horrible decision has to be made we can make it jointly.

### *A SCITT co-ordinator's view of good practice*

School-led initial teacher education will be successful and turn out well-trained reflective practitioners. But I am also confident that it will begin to have a profound effect on those schools involved. The growth of a reflective culture in schools will increase the demand for the service of higher education generally. It is here the real challenge of partnership exists in the future, not in merely improving initial teacher education but working to improve the whole context of education nationally. SCITTs and HEIs should exist in partnership to promote good practice in ITT.

At the heart of the SCITT vision of career-long professional growth is our idea of mentor-driven development; at every point in the CPD process there should be a working dialogue between someone who is an 'expert' and someone who is a 'novice' – we see this idea being formed around our vision of the 21st century of the teacher as a research-driven practitioner.

## TTA guidelines for good practice in SCITTs

From a report prepared by the TTA (1997), good practices within existing SCITT courses were identified as:

- a collective belief in the aims of the course;
- positive attitudes – the will to make it work;
- a commitment to assuring quality;
- a spirit of openness, mutual trust, reflection and critical debate;
- personal and professional support, as well as challenge;
- schools that have the capacity to undertake and sustain a major additional responsibility.

The report also commented on the advantage of the schools being reasonably close together and having a programme of regular meetings to help trainers develop a coherent approach.

Successful groups were shown to share common principles:

- supporting high-quality training;
- achieving their aims cost-effectively;
- sharing the workload rather than concentrating it on a few individuals;
- developing systems that are fair, understandable and supportive to the trainees;
- obtaining the representation of all interested parties, including the trainees.

The structure and management of a course should:

- make clear the legal responsibility for the course;
- assign responsibilities both to committees and individuals for all;

- lay down clear lines of accountability;
- stipulate the mechanisms for ensuring accountability including the nature of regular reports;
- enable thorough quality assurance, by monitoring and evaluation including procedures for actions, where schools fall short of what is required;
- incorporate effective systems of communication;
- enable the course to change and be developed in the light of evaluation and experience.

Questions to ask yourself:

- Do I understand my role and responsibility within the SCITT organization?
- How can I improve my contribution to the SCITT consortium?

Good practice in a SCITT arrangement:

- A school must put professional development at the heart of its practice.
- Its practice must be professional, collegiate, and its expectations must be realistic.
- The mentoring must be open, frank and reflective, and supported by all.
- The vision, culture and language of the SCITT must be shared by all.

# Chapter 11

# Mentoring within a subject department

In this chapter we shall be considering the following aspects of mentoring:

- The role of the subject mentor.
- The role of the subject department.
- Organizing a subject-based mentoring programme.
- Identifying good practice in subject mentoring.

## The role of the subject mentor

The subject mentor has become the key figure in initial teacher education since the swing to an increasingly school-based partnership. This is a key role because of the amount of time the trainee spends with the mentor and because the trainees believe that the place to learn is 'in the classroom'.

There are other key figures in ITT with whom the subject mentor interacts: the professional tutor in school who coordinates ITT activity, and the tutor at the higher education college where ITT is organized in partnership.

The subject-based mentor works alongside the professional tutor and the HEI tutor, ensuring that the various elements of theory and practice are well integrated and balanced in a way that enables and informs the trainee. To separate theory and practice in training would leave the learner, in this case the trainee, likely to fail to make the connection (Gagné and White, 1978). As Nutt and

Abrahams (1994) rightly point out, 'Compartmentalisation makes that which has been learnt on campus poorly accessible in practice.'

Teaching without theory? Whitehead (1989) maintains that all teachers develop their own educational theories in the course of their teaching. Subject teachers cannot just delegate creating theory about how they teach to others. Effective teaching means actively seeking improvement. Your role as subject mentor is to stimulate thought and that means that you are creating a theory of how you teach as you do it.

Your thinking should start with understanding the aspects of initial teacher training programmes and where your departmental contribution fits within it. According to Maynard and Furlong (1993) there are four levels or dimensions of professional training:

A Direct practice. *Practical training through direct experience in schools and classrooms.*

B Indirect practice. *Detached training in practical matters usually conducted within classes or workshops within training institutions.*

C Practical principles. *Critical study of the principles of study and their use.*

D Disciplinary theory. *Critical study of practice and its principles in the light of theory and research.*

As a subject mentor you should take action to contribute at all four levels because as McIntyre and Hagger (1996) say, trainees need access to different forms of professional knowledge – theoretical, practical and a mixture of both – and you can and should facilitate this. The territory of the subject mentor is not to be restricted to the classroom; as a subject mentor you and your department can make an invaluable contribution at the HEI too. Look for opportunities to assist with interviewing, for teaching HE-based sessions and to participate in others, perhaps bringing pupils in to workshop sessions with the trainees in the HE or working alongside the trainees during their in-ICT ITE training. Look also for opportunities to develop your mentoring skills in school across other departments involved in ITT and in discussion with mentors in neighbouring schools and neighbouring LEAs.

Though much has been written about the role of the higher education tutor in the pre-service education of teachers (Furlong and Smith, 1996) and that of the subject mentor (Wilkin and Sankey, 1994, Stephens, 1996) little research has been done about how subject mentoring differs from department to department in just one school. But the nature of the subject is bound to create differences in practice. Mentoring around a content-based National Curriculum like science is inevitably different from mentoring in a subject that has a largely content-free National Curriculum, like modern languages.

Effective mentoring draws from the diversity between subject departments and combines elements of their practice. It seems surprising in a time of 'school improvement' that there is as yet so often little contact between subject mentors across a school. What a great opportunity for sharing!

There is scope for subject mentors to work across schools and this is a model adopted by some ITE partnerships where the school is seen as a mentoring institution. Subject mentors can learn generic skills from one another as well as appreciating different perspectives on approaching mentoring, which would increase their range of mentoring strategies. Time is short in school but it comes down to a question of priority. The school that takes on ITT has two principle foci: the education of the pupils and the education of the trainees. Find time to talk to mentors in other subject areas, lobby for time to do this in school!

As a subject mentor you need to consider what learning to teach your subject means. The following illustration might help you to focus your thoughts about this meaning. This illustration is focused on modern languages but it could apply to other subjects.

---

## Illustration

**(From the University of Bath *Subject Didactics Handbook*, 1999–2000.)**

- How can the study of MFL contribute to pupil learning?
- What is the role of the MFL teacher?
- How can MFL lessons maximize the learning potential of all pupils?
- Which resources are available and which are most appropriate to use?
- How can MFL teachers use the National Curriculum to best advantage?
- How should an MFL teacher assess pupil achievement and attainment?
- What teaching strategies are most likely to result in effective learning?

---

Knowing what 'works' in class is important but as a subject mentor you will also need to develop the ability to explain to the trainee that something works, simply and clearly. This is the mark of a 'good' mentor, who can not only model but can explain.

The subject mentor should prepare for the role by thinking through those aspects of the training programme that are essential to the trainee's development as a professional – and knowing 'that' is only a partial training. The effective subject mentor understands the theoretical aspects of the work of a teacher –

the 'why' as well as the 'that' and can discuss *why* teaching strategies are used to good effect.

Subject teachers know that teaching pupils is more than just telling them. Trainee teachers do not always appreciate the difference and as a subject mentor you will need to reinforce trainees' understanding that learning is active and constructive and requires that they actively participate in learning. The effective teacher elicits these learning activities from pupils in class and shares control over the process of learning with the pupils. As a subject mentor you need to understand what learning your subject as well as what teaching your subject entails.

## The role of the subject department

Ideally, a subject department has ample time for developing a programme of mentoring and making the necessary arrangements to undertake all the preparation. Ideally, the members of the subject department will be consulted well before they undertake ITE and given time and support to understand the ITE subject programme. They should not find themselves under pressure to undertake ITE within their school. In reality, however, the situation is likely to be rather different and there may be little time available, in which case it is even more important to ascribe mentoring roles and responsibilities and make sure the entire department understands the ITE programme before the trainee arrives. Where the subject department finds itself involved in ITE with almost no warning, perhaps when an HE tutor rings the headteacher pleading for a subject placement because the planned one has not materialized, and a student teacher is adrift without a school to go to, it is tempting to agree at very short notice. Such *ad hoc* arrangements may work well, but it is a less than ideal situation for all concerned. Teaching is stressful enough without suddenly adding an ITE loading.

Subject departments should:

- weigh up very carefully the pros and cons of involvement;
- check out exactly what support is offered both by their school and the HEI;
- know that mentoring cannot be marginalized or 'fitted in', because it substantially affects all aspects of the subject department that undertakes it;
- discuss the possibilities that exist for developing and improving their practice by extending reflections from teacher training to encompass all the teachers.

One of the most fundamental changes is that the role of the teachers in a subject department will change to become that of gatekeepers for the teaching profession. Think of the gatekeeper who sits in a draughty, rather cramped glass-sided box at the entrance to a factory, operating the barrier as haulage wagons come and go. The drivers slow, stop, identify themselves and are verified before the gatekeeper allows them passage into the factory compound. Sometimes, the gatekeeper will ring ahead to check that the driver is expected, sometimes the gatekeeper consults a list. Often the gatekeeper's role is to give directions (turn right, park over there) or to ask questions (what exactly are you bringing into this compound?). Less frequently, it is the gatekeeper's role to exclude.

There is a parallel here for the subject mentor. If you are undertaking subject mentoring you will be selecting who will pass into the profession, checking out credentials of a particular candidate for training, giving directions, doing your duty in ensuring that only those who should enter teaching do. Sometimes you will have to turn applicants away. Your subject department will need to support you, especially when things go wrong.

## How does a department assess its potential to engage in ITE?

A willingness to take part is a good starting point – but of itself it is an insufficient base for such engagement. Any subject department considering participation needs to take a long hard look at itself first. Start with a self-audit along the lines suggested below:

- Do we have a suitable mentor?
- Do we have effective management and leadership of our department?
- Do we have appropriate staffing in terms of stability and numbers?
- Do we have appropriate classes for ITE provision?
- Do we have the resources to support ITE (photocopy facilities, textbooks)?
- Do we share a suitable ethos for ITE in our department?
- Do we offer a distinct department style of mentoring?

## Do we have a suitable mentor?

You will find useful guidance in identifying a suitable mentor in the introductory chapter to this handbook but the following illustration, which highlights the benefits of shared mentoring, may give you some further insights into this selection.

## Illustration

In subject department X there are five members of staff – all good practitioners and all with recent experience of working in pre-service training. T is head of department. He is skilled in analyzing why some aspects of teaching work and some do not. He is a highly able assessor. S has an outstanding knowledge of the subject he teaches so well – but is less skilled at explaining why some activities work and others do not. V has recently qualified as a subject teacher. She has a good insight into what pre-service teaching programmes demand. Y qualified four years ago and she is an able teacher. She has excellent interpersonal skills. P is the key mentor – he too qualified within the last five years and his classroom control is outstanding. He is working on improving his interpersonal skills. There are three trainee teachers assigned to the subject department. T works with one, P with two. As the teaching block proceeds it becomes clear that one of the trainees working with P is having problems with classroom management. P, as mentor, has never experienced serious problems and is unable to see why the trainee is becoming distressed. He models good practice and talks through strategies that he finds effective but still the trainee flounders. On the other hand the trainee is learning from him about how to structure classroom activities to reinforce learning. P decides to share the mentoring. The trainee works with Y, who understands her fears, with S to learn effective classroom control and with T as her co-assessor with P. A special mentoring programme is devised for the trainee drawing on the strengths of a group of mentors. Sharing reduces stress!

## Commentary

P is a relatively new mentor. He recognizes that one of his mentees has problems with classroom control and that he is not 'getting through' to her. He knows and trusts his colleagues in the department and calls a meeting with them so that they can share the mentoring and ensure the trainee has support and understanding in the areas where she has problems. P can meets the needs of the trainee with others' help.

## Do we have effective management and leadership of our department?

### *Effective leadership*

A leader has vision and the ability to communicate this to other members of the subject department and draw out their 'best' in meeting this vision. The trainee needs to know not only what teaching a subject entails but how he or she is expected to contribute to this vision while in this school. The subject mentor has the responsibility for getting trainees to know what the department expects of them during their training.

### *Effective management*

The subject mentor needs to enable the trainee to understand how the department is managed and what responsibilities he or she has for sharing in this management. Where, for example, are resources stored and who has responsibility for keeping them in order? Who runs each aspect of the subject curriculum and what are their responsibilities? The trainee needs to attend departmental meetings in order to understand how the teaching across a curricular area depends on an interplay of subject management and leadership. The subject mentor needs to explain this.

## Do we have appropriate staffing in terms of stability and numbers?

For effective mentoring to take place, a trainee needs to appreciate a harmony – a sense of valuing between subject teachers. Learning to teach is difficult enough without getting involved in the friction of micro-politics as a department forms, storms and (hopefully) norms. Effective mentoring is able to draw positively from change but cannot exist in anarchy and infighting between members of a department

The trainee will also need to be valued for having views that may differ from those of the department about what it is to be a teacher. As Dart and Drake say ('Subject perspectives on mentoring' in McIntyre and Hagger, 1996);

> Mentors need to be aware that students' understanding of their subject originates from their own school days. Since the methods they have experienced can often be adopted as the 'right' ones it is a skilled mentor who will enable the student to approach an area with a different and better pedagogy.

## Do we have appropriate classes for ITT provision?

Mentoring means ensuring that trainees have a range of classes to teach and to observe during their school-based placement. A skilled mentor will select with care and balance the experience of the trainee according to their particular needs – consulting the trainee, the tutor and the timetable to look for an appropriate and stimulating mixture. Classes that have been exposed to a number of trainees even where they have been good practitioners need time with their teachers to consolidate their identity and direction, because they can become 'overused' to trainee teaching.

## Do we have the resources to support ITT?

Imagine teaching without resources – no classroom, no board to write on, no photocopying facility, no textbooks, no timetabled sessions. Crazy? Perhaps – but the resourcing needs of mentoring need due attention too. Mentoring is sometimes seen as a bolt-on extra yet it merits more status. Part of recognizing this status is arranging for a suitable place for mentoring and discussion to take place – somewhere comfortable, quiet, private – where you and the mentee can focus and be beyond the call of the pupils. For the mentoring to be effective it needs to have a base – a room – at least for timetabled mentoring sessions for discussion to occur. How are you going to record details of the mentoring process? Is there a board to write on so you can show the trainee what you are explaining, give an outline, provide a visual support to the words that you will say? You should expect to *plan* for effective teaching and for mentoring – and preparation for teaching; and effective mentoring needs resources. Will you use carbon-copy note pads so you can keep a written record of each lesson observation as well as giving one to your mentee and to the professional tutor? Will you use pro formas for lesson planning with the trainee, and will the trainee need to copy those he or she creates?

Do you have access to books on mentoring – in addition to this one – to enable you to learn more and improve your work as a professional mentor?

## Do we share a suitable ethos for ITT in our department?

According to the *Concise Oxford Dictionary*, 'ethos' is the 'characteristic spirit and beliefs of community, people, system, or person'. What characterizes the ethos of your department? How will that ethos be embodied by you, the mentor? You carry a responsibility to ensure that you trainee understands and is integrated into this ethos – while bearing in mind that your school and your department may conflict with the ethos of other schools. That it differs is likely to be useful to the trainee so long as the fundamental values it embodies uphold the very best teaching for pupils in school. There needs to be overt valuing of difference and diversity and a sense of community and commonality of purpose. The best mentoring practice cannot exist in a vacuum but feeds from and into the school ethos. Does your own department reflect this?

## Do we offer a distinct department style of mentoring?

Do you appreciate the difference between being a 'trainer' and a 'teacher-educator'?

Feiman-Nemser poses the question: 'Why don't teachers in mentor-type roles see themselves as teacher-educators?' (Feiman-Nemser, 1998).

Part of the success of good mentoring is seeing it as a valuable experience in itself. Mentors have a direct responsibility for the future of their profession. How they mentor directly affects the future of the profession. How you mentor will affect how your trainees will mentor in their turn. Mentoring is rather like parenting – poor parenting skills in one generation produce subsequent generations of poor parents!

## Organizing a subject-based mentoring programme

Your parenting skills are looking good! You have considered all the aspects of the audit we have been through and you are ready to undertake subject mentoring.

So what comes next? Once you are committed as a subject department to offer ITE get to know your ITE training programme (ask for a copy on computer disc). Understand the different aspects of the course and how your work fits in.

## What to look for in the training documents

- The aims of the course – training courses vary greatly in their ethos!
- The role of the subject mentor – again it varies considerably.
- The role of the HEI tutor if there is one.
- Clear indications of the responsibilities of all concerned in the partnership.
- Contact numbers for rapid access to appropriate people if necessary!
- A statement (that you can agree with) about the role of the subject mentor.

There might also be a recommended teaching loading for trainee teachers – number of hours at different stages in the ITE course, range of ability and year groups.

## Attending mentor development sessions

### *Why bother attending?*

Time is short in school and it may seem too short to justify taking half a day to meet up and chat about mentoring. You need every moment in school because as a mentor you are a teacher and pupils come first! Right? Partly. You have a commitment to educating teachers as well as to educating children. Imagine yourself as your trainee teacher this year. When a mentor does not come for the development sessions a mentee is justified in thinking 'I really don't matter enough for my mentor to attend and learn about helping me to develop my teaching. I'm worthless!'

If the sessions do seem to be a waste of time, be proactive and ask for items to be placed in advance on a written agenda to be circulated several days before sessions.

## So what can be problematic?

Hopefully, there will be relatively few problems in your subject mentoring. By being well prepared along the lines we have considered above you can avoid a great deal of difficulty, but some problems are bound to arise in the course of your mentoring.

---

### Illustration

> I've got a set course, introducing rugby to the kids. Students who aren't specialists are given an insight into being able to teach the game at a basic level. In my first year [as a mentor] we had this huge great fellow who didn't know anything at all about rugby, so he was quite pleased there was [the course] to work on. He was quite confident about taking rugby lessons after that – it was one area he was worried about.

---

Provision for furthering knowledge can and should be made in school as well as in the training institution. Teaching trainees not only *how* to teach within a subject discipline but upgrading their knowledge of the subject itself may be an integral part of the subject mentoring programme. Trainees may not arrive with sufficient subject knowledge and will need assistance in learning how to draw out their pupils' understanding by selective questioning. The problem that trainees encounter in a new area of subject knowledge is that they tend to focus on themselves, but what a rich experience they could have by understanding those aspects of the subject area that are likely to be difficult for the leaner to master – from recent first-hand experience.

As a subject mentor you will need to reinforce trainees' understanding that learning to teach is a lifelong process. In the early stages of training they may well regard you as a teaching icon who has all the answers and all the techniques that teacher might need. While this is flattering and may be partly true it is nonetheless an impression that will inhibit their development if it persists. How should you move them on without disillusioning them or underselling your excellent teaching skills?

- Discuss how you see your subject teaching.
- Identify those aspects of it that need improvement.
- Share your strategies for improving your work.
- Demonstrate that you value co-analysis of practice.
- Be approachable!

Questions to ask yourself:

- What is special about my role in this subject department?
- How can I explain different roles and responsibilities to my trainee?
- How can I show my trainee how to be a good practitioner?
- How can I improve my own subject teaching through mentoring?

Good practice in subject mentoring:

- Carry out an audit to see if your department can support ITE.
- Share the responsibility for subject mentoring with colleagues.
- Ensure that you understand all aspects of the ITE programme.
- Pre-empt problems by careful planning and providing support.

Chapter 12

# Mentoring: a whole-school issue

In this chapter we shall be considering the following aspects of mentoring:

- Why should a school engage in ITT?
- What issues arise when a school becomes involved in ITT?
- Should our school become involved in ITT?
- Identifying good practice in whole school involvement in ITT.

## Why should a school engage in ITT?

> There are two main reasons why schools should concentrate on teacher training. The first relates to factors external to the school: social, economic and political changes. The second is an internal factor: the question of the development of the reflective school.
> (Shaw, 1992)

This chapter begins by touching on the first factor but concentrates on exploring the second factor identified by Shaw – that schools involved in initial teacher training have the potential to develop into 'reflective' schools. For many teachers and HE lecturers, the move to school-based training that followed Kenneth Clarke's speech in 1992 (Clarke was Secretary of State for the Conservative Government) was an overtly political one designed to disempower higher education, based on a deficit model: 'Too often in the past schools involved in teacher training have been left unclear about their role.'

There was little to indicate at this early stage of ITT involvement that so many schools would take on a new role that would empower them as learning institutions, with ITT involvement driving a move to raise standards of teaching and achievement by learners in classrooms. The school improvement movement was in its infancy. If we run the education clock forward to 2000, eight years later, there is an awareness that ITE involvement can empower school development. The future looks promising!

By opening up our teaching to scrutiny within the process of mentoring, we have the basis for examining what we do with a critical eye. It is important to look at why we do what we do and to see whether it is as relevant now as when we started to do it. Is it time to change and are there better ways of doing what we do? Here the mentoring process comes into its own as a two-way dialogue. With trainee teachers empowered not only to attend but to take part in whole-school meetings, they can often bring a valuable perspective to a familiar problem that may not have been possible before. Because they come from a different working context, sometimes they can see a way forward in an objective way, which teachers within the school find difficult to achieve. In effect they become critical friends and their input should be welcomed.

But such empowerment has its costs. Lack of time, lack of resources and to some extent lack of training still dog the progress of the educational growth that teacher training can bring to participating schools. In Clarke's vision, schools would increasingly take over the responsibility for ITE from HE. Mike Berrill and his colleagues at Challney Community School, Luton, where I was privileged to carry out research towards this book, have continued not only to voice but to live out this shared vision (Berrill, 1997). Challney has evolved a reflective school around staff and pupil development, drawing from a passion for improving educational practice, commitment, resilience and sheer hard work from everyone involved. The key lies in putting professional development at the centre.

Why do some schools still not see ITT as the focus of their school development? When you ask headteachers this question the usual response relates to how they and the school community see their hierarchy of educational priorities. Initial teacher training is recognized as being important but not as important as educating pupils in classrooms, and specifically not as important as raising scores in government league tables for public examinations. 'A man cannot serve two masters', so the saying goes, but when two can enrich each other they form a unity greater than the sum of the parts. So it is when a school takes on a genuine commitment to the education it should see initial teacher training as an integral element in this. The Framework for Professional Development, the national standards under development by the Teacher Training Agency (1998), reinforces this integration of ITT .

## What issues arise when a school becomes involved in ITE?

Having established that involvement in initial and continuing teacher education is likely to enhance the educational potential of schools, it is time to look in detail at the issues that such involvement raises for individual schools and how theses can best be managed to benefit all. Watkins and Whalley (1993) set out a clear and comprehensive list of the hallmarks of effective initial training of teachers:

- It is an active process of learning, involving a number of parties.
- It is based on practical experience.
- It aims to support beginner teachers in reflecting on their job and their development into it.
- It is a developmental process both in term of the period of initial teacher training and in terms of linking to future professional development.
- It recognizes that teaching is a complex professional activity.

Watkins and Whalley usefully examine the main issues that schools involved or considering involvement in ITE should consider. Having begun with an exploration of the context and a rationale for school involvement in ITE, this section of the chapter will consider these issues, which raise a host of questions.

What will be the likely effects of school involvement in ITE:

- upon classes?
- upon individual children?
- upon mentoring staff?
- upon non-mentoring staff?
- upon support staff?
- upon the school community as a whole?

As Shaw (1992) so rightly says, the first priority for a school contemplating involvement in ITE is to carry out a self-audit. Self-audits seem to be out of favour at present with OFSTED, apparently shunning the pre-inspection self-consideration that many schools have undertaken. Yet without an audit how can a headteacher be sure that everyone has had a fair chance to have their say and to offer their expertise or point out their need for professional development? Shaw's audit is in two parts, with one questionnaire for trainees/new staff and another for middle managers. Her questions delve beyond the surface organization of ITE and explore opinions.

This is an essential perspective for identifying whether a school can undertake ITE:

- Do you feel confident about assuming responsibility for trainees?
- What in your opinion were the strengths of school induction?
- What in your opinion were its weaknesses?

This is in addition to, and not instead of, systematic enquiry into the organization of the training programme:

- What documentation does your cluster, department or year give to trainees?
- What documentation does your cluster, department or year give to new staff?
- What elements could usefully be added to departmental induction?

A sense of ownership emerges as all staff are invited to contribute to the pool of knowledge, opinion and experience that grows as responses to the questionnaires are collated. Involvement in ITE is something that develops across the school and not from a top-down imposition by the senior management team. Strength lies in collaboration.

Establishing who can offer what, who wants to be involved and who can be supported in the first and subsequent waves of ITE staff development is an exciting but exacting challenge. It is a process that will raise a number of issues in the process.

These include:

- communication;
- dealing with challenges and conflicts;
- resources;
- management of the learning experience;
- multiple mentors.

## Communication

Clear lines of communication are essential to effective practice in ITE but pre-ITE involvement practices may leave staff feeling out on a limb when the school undertakes a commitment to train teachers. Who is in charge of staff development? Should the same person have overall responsibility for running ITE?

If your school is considering the twin foci of induction programmes for new staff, managing and delivering in-school continuing professional development:

- Is the role to be divided between two or more people?
- How will they communicate on a regular and systematic basis?

- When will they meet?
- Is everyone in the school familiar with the language of ITT?
- Does everyone realize the implications of involvement in ITE?

Communication requires the parties to be prepared to listen as well as be ready to speak out.

Does the school have sufficient established lines of communication that can deal effectively with outside agencies contacting departmental mentors and still retain integrity in its operation and the quality of its participation across several departments? How is information to be centralised and how fed out from the centre?

## Challenges and conflicts

Undertaking involvement in ITE brings its own challenges and conflicts to a school community in addition to those that face schools daily. Difficulties arise most frequently when trainee teachers encounter problems in classroom management, in subject and pedagogical knowledge and in adjusting to the pervading culture and traditions of the host school. Challenges include the sheer amount of documentation involved in ITE, staff turnover mid-year, particularly among mentors, and handling funding. Asking how funding attached to ITE should be allocated in school is likely to be a contentious issue, as can allocating ITT responsibility to just some departments. In preparation for discussing ITE allocation roles, consider the following questions:

- Who decides how funding will be allocated?
- Who receives funding?
- Does funding go to pay mentors directly or for departmental resources?
- Do mentors get to rotate their role?
- Do departments get time out of training to give staff and pupils a rest from training new teachers?

## Resources

It can be the small questions that make the difference between a programme that succeeds or fails:

- How will the school deal with trainees' requests for photocopying?
- Will they be allocated a key number for the photocopier?
- Or will they have to place requests for copying via their mentor?

The nitty-gritty of involvement needs to be sorted out in advance. There need to be policy meetings for teaching staff and also for ancillary non-teaching staff, otherwise chaos will soon ensue! Policies need to be clear and agreed if time is not to be wasted checking and rechecking and redressing disaster. Again clear lines of communication are needed to get a decision if an unforeseen necessity arises. You need someone with an overview and with time and patience to solve challenges before they escalate.

Even the locational aspects of involvement in ITE are potential problems:

- Where will the trainee teachers sit in the staffroom?
- Where will they eat?
- Where will they take a break?
- Where will mentoring sessions be held?

## Management of the learning experience

Schools intending to become involved in ITE need good managers and teachers. Management of ITE involvement is demanding and encompasses many questions:

- Is the school familiar with the demands of the training programme?
- Can the constituent parts be made to fit together comprehensively?
- When will mentoring sessions take place?
- Who will take classes in an emergency when the mentoring time is protected?
- How will the quality of mentoring be monitored in the school?
- How will new mentors be trained and supported within the school?
- Can staff deliver the general educational, the pastoral and subject didactic elements?
- How will learning by trainee teachers be managed in school?
- Are suitable classes available?
- Is there time and staffing to keep a regular programme of lesson planning observation, and debriefing within the confines of the school day?
- How many departments would like to train teachers?
- Are they suitably staffed?
- Can the school manage an influx of several new staff in a short time?

Good management is about effective delegation and it should be obvious from the questions raised above that no one person can adequately address every ITE issue. There needs to be a coordinated approach to allow for input from several 'experts', and collaborative management, which requires the appointment of suitable personnel and a realistic time allowance for ITE work.

## Multiple mentors

In addition to subject mentors the school will need a general professional tutor who oversees the mentors and coordinates the school ITE policy:

- Is such a person available and how much training is he or she likely to need?
- Is one mentor sufficient per subject area?
- Or would it be possible to have one lead mentor and another under training?
- When mentors have received training at the HE institution will they be enabled to pass on their expertise in mentoring to colleagues in school?

## Learning from experiencing ITE involvement

The decision to undertake a commitment to ITE is agreed among the staff. Not everyone is on board fully but there is no overt opposition and everyone is prepared to 'give it a go' for a limited time. It is important to plan for debriefing and review:

- What are the mechanisms by which the school will feed back what it learns about mentoring to itself – to improve future practice in school – and to others such as other schools and members of the teaching staff at the partner institution?
- How will the school know whether it is being successful?
- How will it evaluate its progress? Will it seek feedback from everyone concerned before initial steps are taken in undertaking teacher training?
- How will good practice be disseminated and how will it be celebrated?

Before partnership or SCITT involvement in ITE is begun, these questions are waiting to be answered fully, openly and convincingly if there is to be effective mentoring in a whole-school coordinated way. The level of the school's involvement needs to be negotiated – will the school take trainees from just one HE institution or several? Which HEI? All subject departments, or would it be best to start with one?

## Should our school become involved in teacher training?

You can find an overview of the benefits and drawbacks of ITE involvement in chapter 1 of this handbook. Once a school has carefully weighed up the advantages and disadvantages of potential involvement in ITE and has made a commitment, planning begins! Whatever model of ITE involvement is undertaken the school community will be inexorably changed as a whole. If this change is to be for the better it requires meticulous planning. All members of the teaching and non-teaching staff need a sound awareness of the issues involved before any trainee teachers arrive, before any contract is signed, any potential trainee called for interview. Schools need to weigh up very carefully their own suitability for ITE. There will need to be not only liaison about organization of ITE but agreement about how this will be managed and by whom.

There will be a need for on-going discussion as the process of involvement in ITE unfolds, as we can see in the following illustration and the question it poses.

---

## Illustration

A school has been in partnership with a local HE for five years. The departmental links with two subject areas have been highly successful and the PGCE students have reported that the school placement has met their needs. There is an OFSTED inspection. The outcome is depressing. Although one department is rated as excellent and the other as good, serious weaknesses are identified in other subject areas. A second inspection shortly follows the first. A similar report ensues. *Should the school maintain its involvement in ITE at the present time?*

Involvement in ITE will never be plain sailing. There will always be difficult transitions and decision making in its day-to-day organization. If we put consideration relating to arrangements on one side for a moment, let us consider the reason for a school to exist. Surely it is to be a centre, a focus, a crucible for community and learning. A haven and a nurturing ground for developing life skills and academic potential – an environment where education means drawing out as well as feeding in learning in a compassionate and knowing way – where creativity and imagination are valued for their contribution for the good of all. Put this ethos at the heart of a school and it is difficult to see how there cannot be involvement in teacher development in a career-long continuum.

Undertaking ITE requires sound planning and good management. Some issues for schools to manage prior to involvement include:

- developing whole-school awareness about possible involvement;
- raising staff awareness;
- defining the role of SMT in setting up involvement in ITE;

- defining the role of subject departments in setting up involvement in ITE;
- ensuring the pastoral organization is prepared for involvement in ITE;
- putting ITE at the heart of the school mission statement and development plan.

Good practice is a whole-school issue – it doesn't just 'happen' by itself!

## Illustration

The following are extracts from a taped interview with a headteacher:

> I don't know that I can list the advantages. I think it's an ethos, if you like. A feeling. There are great advantages for the departments because it gives them an opportunity to speak with someone who's recently been trained, they're given ideas. The student teachers are great ambassadors for us, they really are – especially when we went through a bad phase in the local press over an excluded pupil. The students took the time to write a letter saying how much they enjoyed their time here – they didn't say there was no bullying but they did say it was dealt with.
>
> Mentoring would certainly go on (my staff's) CV and any reference. You see, the people who've mentored never apply for jobs. They stay. It's not actually on anyone's job spec apart from the senior school tutor's. I wouldn't want to put it on anyone else's really, because if you've got a job spec you're meant to work to that job spec and I like the opportunity that if someone says 'Look I've done it for two years. Why not let so and so have a go. X will be ideal next year, I've got a new lady joining the science department and she'd be good.' So to rigidly tie it down is no good. It's a voluntary thing, if you like, they're doing, but although it's done in a voluntary capacity I think they're still doing it really well!

## Commentary

In a sense this illustration speaks for itself. An inner-city school with more than its share of social and economic problems. School attendance is good and so is staff retention. Trainee teachers continue to vie for placements there each year...

Putting ITE at the heart of a school's activity means ensuring that it is included in the school development plan. Just how to do this will be a matter of individual choice, but the following illustration shows how one headteacher has approached this.

## Illustration

> ITE is written into the school development plan. This is a school improvement plan and we've got six broad priority areas for school development which are whole-school ones:
>
> - values and behaviour;
> - improved learning;
> - ICT;
> - environment;
> - personnel;
> - self-evaluation.
>
> ITE comes into all of those, especially under self-evaluation.
>
> . . . [we use the funding from ITE] to buy in someone to give us advice about school improvement and we say 'Let's make sure we are using our trainee teachers in that respect too.' In my previous job I did quite a lot of management development work. I do quite a bit in school trying to develop people as reflective practitioners, developing colleagues with management competencies so they stop and think about what they manage and how they manage, and some of the generic skills apply to heads, deputies, heads of department, heads of year, NQTs and would-be teachers. There are three strands to the school I'm trying to develop here. One is the notion of the learning school. Another is the notion of the self-evaluative school, reflecting on what we do. The third aspect is the notion of the assertive school, which means that I want everyone to have the right to have a say.
>
> I spend a lot of time thinking about leadership and accountability in schools because I have to take the lead and I'm accountable, but my theory and belief is that everyone in school from the lowliest teacher to the lowliest pupil is actually a leader.

## Commentary

In the short time since this headteacher was appointed the school has become a technology college and been awarded two prestigious awards for its work in actively promoting community among its pupils and its staff. Whole-school involvement in ITE means moving together in a reflective valuing and coordinated way. Initial teacher training is not a stand-alone or a bolt-on extra to the daily operations of the community. It is a natural focus within a learning community. Mentoring becomes part of the fabric of the school. Think about mentoring in terms of continuity, building on what has happened in a trainee's pre-service education and linking it via the career entry profile (CEP) to the induction year. Look for opportunities for teachers to mentor other teachers across and within departments and share their expertise freely. Like this, constructive criticism can be offered and taken on board in a positive way. A pipe dream? Make it a reality! It takes time, dedication and foresight to achieve, and the right staff.

(You can find out more about Learning Schools from Holly and Southworth, 1989.)

---

Questions to ask yourself:

- Is ITE really a whole-school issue here in my school?
- How can I help to make ITE a whole school issue?

Good practice in whole-school involvement in ITE:

- Explore generic issues surrounding involvement before undertaking ITE.
- Explore issues relating to your whole school before undertaking ITE.
- Ensure that you monitor and evaluate ITE involvement on a regular basis.
- Look to link pre-service and induction-year experiences through the CEP.
- Maximize the potential of putting mentoring at the core of school practice.

# Part 4. CONTEXTS FOR MENTORING

Chapter 13

# Form tutor mentoring

In this chapter we shall consider the following aspects of mentoring:

- The need for training in form tutoring.
- Activities to promote understanding of form tutoring.
- Creating a programme for form tutor mentoring.
- Identifying good practice in form tutor mentoring.

## The need for training in form tutoring

> No-one would deny that student teachers should be prepared for pastoral work in schools. Preparing student teachers for pastoral work has traditionally been one of the more problematic and neglected areas of pre-service education. With school-based ITE there is the opportunity to deal much more thoroughly and effectively with this aspect of teachers' work and to make sure that student teachers acquire a working knowledge of their pastoral responsibilities as teachers.
> (McIntyre, Hagger and Burn, 1994)

Teachers in Britain have traditionally prided themselves on their concern for the 'whole child'. This has attracted international attention as a central feature of primary schools following the publication of the Plowden Report (Department of Education and Science, 1967), and is also reflected in the pastoral care networks of secondary schools. It contrasts with the narrow curricular focus of some other countries in Europe. French teachers, for example, are reported as being amazed by the range of tasks that their English counterparts take for granted. One of the difficulties for new entrants to the teaching profession is that their more experienced colleagues often appear to equate pastoral care with discipline, organization and routine. Tutor periods are, in some schools, little more than a time to collect sickness notes and to distribute letters to parents. How can this be equated with the notion of educating the 'whole child'? As Wootton (1992) says:

> Being a form tutor has, in some respects, elements of parenthood and friendship: form tutors frequently speak of children and they try to relate to them in a different, more personal, way.

It is so important that new recruits to the teaching profession are trained to avoid overfamiliarity: to recognize that theirs is a professional relationship and that they cannot really imitate the personal relationship of family and friends. They need to learn what is a discreet professional distance and avoid becoming personally involved in family matters.

Skills set out as a requirement for meeting the present government's standards for Qualified Teacher Status only indirectly link the novice teachers to effective form tutoring. There is no section of the standards dedicated to this important role, although there should be.

However, the 1995 edition of the Inspection Handbook for OFSTED (page 89) does stress the centrality of the role of the form tutor:

> In many schools the responsibility which the class tutor carries can facilitate a coherent approach to pupils' spiritual, moral, social and cultural (SMSC) development and its connections with intellectual development . . . Inspectors should evaluate the extent to which provision (for pupils' SMSC development) reaches all pupils whatever their background.

Given this focus by OFSTED, it is clearly desirable for novice teachers to gain experience in form tutoring by working alongside colleagues in schools. With the advent of school-based teacher training a new body of teacher-educators (mentors) emerged. They were normally subject specialists in the secondary sector, although in the middle and primary sectors the division between subject specialist and form tutor was necessarily less distinct. The challenge facing providers of initial teacher education was to ensure some kind of parity of experience among novice teachers during their school placements. Novices are to be prepared to teach in a specific sector of education – and their experience of working in schools during their pre-service training must be suitably broad to equip them to teach in schools other than the ones where they train.

In fact this has been less problematic than at first appeared, from a pastoral point of view at least. Most schools operate traditional horizontal structures within the pastoral system. Most schools operate around year groups rather than vertical house groups. Most schools also offer a generally positive perception of pastoral care and increasing interest is being shown as we move towards a major review of the National Curriculum that includes more overt and defined ways of ensuring pupils' welfare.

## Activities to promote understanding of form tutoring

New entrants to teaching need preparation for form tutoring, a role that has traditionally received too little attention in initial training programmes and has left many experienced teachers unsure of their ground as form tutors. As Marland (1974: 12) points out: 'the proportion of competent teachers succeeding as Tutors is lower than that of those succeeding well in their subjects.'

If you are preparing to mentor an NQT you may find that he or she still has relatively little understanding about how the form tutor role works. There may have been limited opportunities in previous training to assist a form tutor but taking on the sole responsibility for an entire tutor group for the year is a whole new ball game! If the teaching profession is to address its own shortfalls, a systematic programme of initial and continuing professional development for form tutors really is long overdue. In your mentoring you can do much to ensure that this gets under way.

All teachers, and therefore form tutors, have legally binding contractual responsibilities and statutory duties. Section I of the 1988 Education Reform Act requires schools to undertake a balanced and broadly based curriculum which should promote the spiritual, moral, cultural, mental and physical development of pupils. There is a substantial overlap in the responsibility of teachers and tutors to 'facilitate a coherent approach to pupils' spiritual, moral, social and cultural development'. Is it possible to identify differences between the roles of form tutor and subject teacher? Griffiths, in Best, Lang, Lodge and Watkins (1995), sees a substantial overlap and, similarly, Griffiths and Sherman (1991: 17) conclude that:

> The roles of the form tutor and the teacher do appear to us to be different in some respects though not . . . in terms of the skills applied or even the tasks undertaken. One of the differences lies in the area of autonomy or, conversely, of boundaries.

Griffiths and Sherman usefully set out the relative importance of several elements as significant components of either a form tutor's or a subject teacher's role, or both. In their survey of teachers' and tutors' responses to a questionnaire designed to delineate relative responsibilities, the significant differences lay in three areas where the form tutor carries more responsibility:

- To record on the pupils' files all relevant information, keeping these files up to date and ensuring that senior colleagues are aware of any changes.
- To ensure that letters to parents are delivered, timetables and other routine administrative matters are dealt with as speedily as possible and that, as appropriate, notices of detentions, medicals, special meetings and so forth are passed on.
- To complete the termly assessments to parents, also coordinating other teachers' reports and giving a coherent comment on the 'total pupil'.

Some teachers, often well qualified academically, still fail to understand the essential point of pastoral care. They see themselves exclusively as subject specialists. They see 'the pastoral staff' as crisis managers employed to remove tiresome problems and to 'deal with' such alarming matters as physical and sexual abuse, drug-taking, or family breakdown.

Those who think like this will never become really effective teachers. Theirs is a narrow and inefficient view of the teacher's role. It is simply not possible to teach effectively without taking account of the child's personal motivation; and this in turn means that you must be prepared to recognize, and if necessary spend time on, those features of personality and behaviour that aid or hinder the pupil's learning. Most of these concerns involve aspects of behaviour, school attendance, relationships with other pupils and staff and attitudes to work.

Reynolds (1995) reporting on form tutoring activities in a mixed comprehensive for pupils aged 12–18, found that form tutors held for the most part an individual view about the role, seeing aspects of casework among their more important tasks. Group work appeared to occupy a subordinate position in their concept of the role. At group level, the major task was seen as promoting a good atmosphere and sense of coherence. The National Curriculum was seen as having a significant and negative impact on time available for teachers to fulfil the tutor role. The competitive pressures generated by the 1988 Education Act were felt to be having an effect. Tutors felt greater expectations upon them with regard to attendance, uniform and appearance. Greater parental contact was taking place through year heads and senior management.

## Expectations of form tutoring

Many who join the PGCE course have no experience as pupils of the form tutor role, particularly those from abroad or those whose experiences of form tutors as pupils were minimal:

> The role of the form tutor is one of the most nebulous concepts facing the novice teacher, encompassing a range of duties from the dry fulfilment

of legislative commitments to intuitive understanding and empathy towards a group of very different individuals.
(PGCE student, 1998)

Some novices may need an induction period in school to explain the responsibilities that are part of a form tutor's role and the expectations your school has of its novice teachers involved in this aspect of their practice of teaching in the PGCE course.

These questions have been compiled from inquiries by a group of novice teachers about their role as prospective form tutors. Clearly some of the novice teachers had little conception of the role:

- What is the main role of the form tutor?
- How much non-timetabled time is involved in form tutoring?
- What happens in tutor group time?
- What does personal and social education (PSE) entail?
- Is the form tutor necessarily responsible for teaching PSE?
- How close, emotionally, should a form tutor get to their group?
- Are there any taboo issues that should not be discussed with your class?
- How do you ensure that more able and better behaved pupils are not ignored while pupils with problems are getting attention?
- Is the role of the form tutor more one of facilitator than instructor?
- How do you get to know your pupils when time for one-to-one contact is so limited?
- How do you deal with children who fall behind in certain subjects?
- Is there a handbook to advise form tutors on how to act if a child in their class is unhappy or has personal problems?
- What sanctions and rewards are available to form tutors?
- What are the pupils' perceptions of tutor time?
- How does the use of tutor time vary between year groups?
- How do you prepare reports on pupils' work to send to parents?
- What are the qualities of an effective form tutor?

So much for the burning questions posed by trainees about form tutoring, but what are your expectations of your trainee as a form tutor?

## Form tutors' expectations of novice teachers

Some teachers are keenly aware of the need to work alongside trainees as they get to grips with the form tutoring role but others seem prepared to hand over control, with little support. A group of teachers in the Bath area drew up a list of questions that they felt would raise their own awareness of the particular needs of the novice teachers with whom they were to work.

### *Issues for planning and discussion (prior to meeting the novice teacher) between the form tutor mentor and the SMT*

- What are our own expectations of a novice teacher?
- How much does our novice teacher know already?
- What does our novice teacher need to know?
- What kinds of experience might we need to plan?
- How do we plan our mentoring cycle in terms of planning, doing/observing, giving feedback?
- How much support should we give our novice teacher?
- How do we ensure a suitable balance of support and challenge?
- What kind of support might form tutors need in mentoring?
- Where will form tutors obtain that advice and support in school?

## Creating a programme for form tutor mentoring

These are useful activities for trainees to undertake alongside the form tutor:

- completing the register;
- helping with assemblies;
- organizing/keeping a tutorial notice board;
- attending year group meetings;
- observing/supporting the form tutor at parents' evenings;
- welcoming new pupils;
- supporting individual pupils (for example, on choice of courses, homework diaries, homework, specific problems);
- giving out notices;
- supporting groups or individual pupils in PSE.

### *Getting discussion going*

The following activities relating to form tutoring can form the basis for useful discussion between form tutors and the pre-service or newly qualified teachers assigned to them:

- counselling and tutoring pupils;
- observation and exploring changes in patterns of attendance and behaviour;
- understanding causes of pupils' success and failure;
- supporting records of achievement and profiling;
- understanding family/ethnic and cultural experiences of pupils;
- communicating effectively with parents/guardians;

- linking with and making use of the local community as a resource;
- communicating effectively with colleagues in the pastoral team;
- contributing to whole-school review and curriculum planning;
- making appropriate referrals in school, in wider welfare work;
- understanding educational/training and welfare networks;
- professional self-review and development in pastoral care.

(List adapted from NAPCE, 1991)

## Issues to consider in relation to mentoring

- What are the assumptions that your novice teacher brings to form tutor work?
- What are the similarities and differences in the role of the form tutor between the trainee's experience and your situation at school?
- Has the novice teacher a clear understanding of the pastoral system at your school?
- Is there, perhaps, an opportunity for several of the form tutors to each spend a little time with the group of novice teachers about their work?
- Has your novice teacher had the opportunity to see that sometimes there is no one right way of handling a situation, that sometimes 'least worst' is best solution!
- What kind of support do you, as a form tutor, need to prepare you for and support you in your work with the novice teacher?
- What kind of support is currently available to prepare you for and support you in your mentoring role with novice teachers at your school?

## Questions to ask your novice teacher

Think about your own schooling: how far do you think your teachers were concerned about/responsible for your personal and academic welfare?

Consider, as a teacher, how far do you think it might be reasonable for the school and parents to expect you to act *in loco parentis*?

So far we have looked at the need for mentors to have guidance and support for their mentoring but for form tutoring, especially in secondary schools, there is a need for training for other staff involved in the mentoring of trainees: the form tutor-mentors.

## Illustration

The following list of guidelines for practice within the PGCE was developed as a result of discussions with staff at the University of Bath and one of our partnership schools.

A preliminary meeting should take place in the school before the beginning of the PGCE term, between the professional tutor, the education and professional studies tutor and all form tutors who will be working with the novice teachers. Form tutors can use this opportunity to find out about individual trainees who will be working with them and to discuss the professional expectations of their role.

A second meeting is expected to take place in school at the start of the first serial phase between the professional tutor, the education and professional studies tutor and all form tutors. This will provide an opportunity to ensure that all trainees are aware of the expectations of the whole-school aspects of the work they will be engaged in.

In addition, the form tutor mentors should have the opportunity to meet the professional tutor in school time at regular intervals during the year to discuss the progress of trainees.

Crucially, the form tutor mentor should be consulted in the eventual appraisal of the trainee.

## Commentary

The tutors at this school realized before they undertook mentoring that they had a vital role to play in the development of novice teachers' awareness of and skills in being a form tutor. They took up the opportunity offered by the deputy headteacher to assume a proactive stance. The school planned a series of meetings with the university with whom they worked in partnership within the PGCE course and a together they drew up a list of priorities in mentoring. These included a series of targets to be implemented before the novice teachers arrived, on their first meeting and during mentoring sessions. These targets were translated into action plans for the university and for the school and have formed the basis of the University of Bath's *Form Tutor Digest*. Before this proactive stance of the school the PGCE course contained little by way of overt training in form tutoring and there was almost no overt support for tutors undertaking mentoring in school. All schools in the PGCE partnership are now issued with copies of the *Digest* prior to the start of each academic year. This forms part of a dedicated programme of tasks, monitoring and assessment of novice teachers' progress.

---

Looking at the organization of the form tutor programme is an essential aspect of good practice in mentoring but there is more to consider than just organization, as the next illustration shows.

---

## Illustration

The mentoring programme in this school focuses on the attributes of a successful form tutor rather than on the organization of tutoring as a starting point for a mentoring programme. It poses questions for the trainee.

Are you:

- Organized? (Paperwork under control and meeting deadlines?)
- Approachable? (Never too busy or too rushed for them to approach you with a problem?)
- Diplomatic? (Dealing with the situation itself rather than 'taking sides'.)
- Genuine? (Proving you are worthy of the class's trust.)
- Conscientious? (Giving tutor periods status: not just an opportunity to catch up on marking!)
- A team builder? (Thinking about 'we' rather than 'you' when it comes to activities in school.)
- Likeable? (But not ingratiating! Don't expect all students to like you . . .)
- Interested? (But not too inquisitive – finding out about your students by talking not 'grilling'!)
- Fair? (Aware of personal prejudices and working with the conviction that all pupils have a right to be offered the same level of consideration and courtesy by their teacher.)
- Contented? (Looking for ways of enjoying being with your tutor group.)

## Commentary

Form tutoring is about people not just about procedures involving people. It is relatively straightforward to ensure that trainee teachers work through a list of pre-determined tasks relating to form tutoring – marking the register, collecting sick notes, etc. However, if the teaching profession is to redress the imbalance highlighted by Marland in the introduction to this chapter that the proportion of competent teachers succeeding as tutors is lower than that of those succeeding well in their subjects', all form tutor mentors should seek to ensure that their mentees are aware not only of the mechanics of form tutoring but of the personal qualities that distinguish an adequate from an expert form tutor. Once aware, the novices set about developing similar attributes to these.

---

Back to practical considerations! How do you enable your trainee to grasp the intricacy and pleasure of well-managed form tutoring? You can begin by setting a number of activities that will bring your trainee into contact with different personnel involved in form tutoring.

Ask your trainees to report back to you once they have answered these questions:

- What is a typical daily routine for a group tutor?
- How much time is spent on this role?
- How different is it from the teaching role?
- Are different skills needed, or is it just the context which is different?
- Who is responsible for the coordination of the work of group tutors?
- Is there a common programme across the school or the year group?
- How much responsibility does the individual tutor have for devising a programme?
- How and under what circumstances does the tutor interact with individual pupils – are there two or three examples of this which will give an indication of how different the tutoring role is from that of the teacher?
- How, and when, do tutors interact with subject teachers?
- Do group tutors meet parents on a regular basis or only occasionally, for example on parents' evenings?

Once your trainees have an understanding of what form tutoring entails it is time to help them develop their skills through taking on (limited) responsibility. It is *not* time to hand over the tutor group to trainees and to say 'Well, off you go then. You'll soon get the hang of it!'

Your trainees' future and their pupils' wellbeing depends on your doing more.

Questions to ask yourself:

- What do I know about mentoring and where can I find out more?
- How far am I willing and able to be a mentor for form tutoring?
- What does form tutoring entail as far as I am concerned?
- What do I need to know about the ITE programme and where I fit into it?
- What are the particular needs of my trainee form tutor?
- How am I going to meet those needs?
- How am I going to plan, observe and give feedback in a professional way?
- Where is my opportunity to give feedback to the ITE provider on my trainee?
- What are my own values in relation to form tutoring?
- What are the areas of form tutoring that I know I should improve?
- How am I going to improve what I do and how can I enable my trainee to help me?

Good practice in form tutor mentoring:

- Ensure that your trainee sees the scope of form tutoring across your school.
- Provide the opportunity to share mentoring with experienced tutors.
- Encourage your trainee to develop personal as well as organisational skills in tutoring.
- Help your trainee to appreciate just how fun and useful tutoring can be!

Chapter 14

# Using the Career Entry Profile

In this chapter we shall be considering the following aspects of mentoring:

- Requirements of the Career Entry Profile (CEP).
- Ensuring continuity between pre-service and NQT stages in ITT.
- Creating and making good use of a professional development plan.
- Identifying good practice in using the CEP.

## Requirements of the CEP

In 1995, the Teacher Training Agency produced a pilot career entry profile scheme for teachers qualifying in England in 1996. There was widespread support for the principle of career entry profiling and CEPs were seen to have the potential to help students/NQTs take responsibility for their own development and to provide a valuable springboard for discussion relating to induction and other aspects of professional development (page ii).

### *The format of the CEP; current statutory requirements*

There are four sections; A, B, C and D:

- Section A is a summary of the NQTs' initial teacher training.
- Section B lists the NQTs' strengths and priorities for their further professional development during induction.

- Section C gives the NQTs' own targets for the induction period.
- Section D sets out the targets and action plan for the induction period.

Section B requires agreement between the ITE provider and the NQT and section D between the school and the NQT.

## The CEP year: an overview

The CEP process begins in the last year of ITE and spans the first year of teaching.

**April to May:** for September-start ITE courses, the CEP arrives on the scene.
**May and June:** it is the responsibility of the ITE provider and the trainee to complete the first two sections (A and B).
**June and July:** once QTS is formally recommended, the ITE provider reviews these sections, signs the profile, keeps a copy and sends the original to the NQT.
**June to August:** NQTs consider priorities for objective setting and action planning in preparation for developing their action plan once in their first post.
**September:** the NQT and the induction tutor complete section C.
**November to June:** NQTs and their induction tutors review progress and revise objectives in the profile.

## Ensuring continuity between pre-service and NQT stages in ITE:

In this chapter we shall be considering sections B and C, where negotiation is required between the NQT and the ITE provider and later between the NQT and the school.

The CEP thus forms a bridge between the first and second phases of the process of initial teacher education, between pre- and post- qualification stages. If the foundations of the bridge lie in each stage and spans extend towards one another then it is down to the NQT with mentoring assistance to join the two. The NQT and the mentor need support for their work from the school as a whole.

Accurately marrying the two sections of the bridge is a highly skilful job, not to say a precarious one, and cannot to be rushed or overlooked. The alignment must be exact and this joining is achieved by skilful manoeuvring and welding. In the first span the school-based mentor and where appropriate the HE tutor have a responsibility to prepare the trainee for identifying the targets that will be translated into action plans. In the second span it is the induction tutor's role with coordinated assistance from members of the school's senior management team. The mentoring from the first stage should inform the work of the induction tutor and the SMT in the second.

Clearly the NQT needs as much support as possible from both providers. So, too, do the mentors who need space and time to make a professional job of linking both sides of the training programme by setting out the objectives and action plan for the induction period. This is a considerable responsibility and as the guidance notes that accompany the CEP (1999: 8) say:

> The way in which the objectives are framed will affect how achievable they are, and the ease with which progress towards them can be monitored and reviewed. Objectives should be realistic and attainable.

There is a weighty responsibility lying on the shoulders of the NQT, who needs to be clear about his/her needs because:

> new teachers should take their profile into their first teaching post and, working with their induction tutor, use section C to agree and record objectives for professional development and a related action plan for the induction period.

For the CEP to work as intended, the stages in it should form a seamless continuum.

If this seamlessness is to be achieved the trainee needs experience in managing effective target setting and action planning in the period before attaining QTS.

## Creating and making good use of a professional development plan

In the next illustration we shall see how one provider of pre-service training is managing this by requiring trainees to create a professional development plan.

---

### Illustration

The professional development plan (PDP) has been in operation at the University of Bath for a number of years and is an integral part of the PGCE course for both secondary and middle years pre-service teachers. It is used by the trainees to provide a record of their progress through the year in respect of the standards for QTS and general competence in relation to their school and university-based work. The PDP is the main framework for monitoring and assessment of progress that leads to the completion of the CEP in addition to the practice of teaching reports completed at the end of each of the three blocks of teaching in schools. It provides a link between the trainee, the HEI, the school, the standards and results in the CEP.

The main principles and advantages of introducing a PDP can be summarized as follows:

- It is a self-evaluative tool for the trainee.
- It is a focus for recording reflective practice with goals and action plans.
- The trainee has ownership of the process.
- It develops a shared language for discussion with mentors and tutors.
- It develops a shared culture (for improvement) of teaching.
- It develops the vision of teaching as artistry.

The PDP is divided into two main parts.

In the first part the PDP divides the standards for QTS into five sections, dividing the second area of the standards for QTS into two more interrelated but more manageable sections:

- Section A: subject knowledge and understanding.
- Section B: planning.
- Section C: teaching and class management.
- Section D: monitoring, assessment, recording, reporting and accountability.
- Section E: other professional requirements.

For each of the sections there is a 'slidable' grid where trainees and mentors can plot progress at regular occasions over the PGCE along a continuum which records 'not started yet', and 'making sound progress', towards 'consistently demonstrated/achieved'.

For each of the five main sections of standards, trainees are expected to give a brief description of the evidence they are submitting to demonstrate their achievement of individual competences. This evidence is cross-referenced to a variety of sources, which include lesson plans, assignments, notes made during mentoring sessions, feedback by observers on lessons taught, and notes made during university-based teaching sessions. Each piece of evidence submitted is logged with the date of entry.

In the second part, trainees are provided with four sets of pro formas where they review their progress, noting down their strengths and setting priorities for development. As the grids in the first part of the PDP only provide a listing of the evidence against the standards, the four double-page review sheets in the second part are for the trainee to give an overview of progress already made and an opportunity to identify priorities to concentrate on. These double-page review sheets are similar to the ones in the practice of teaching reports completed three times a year and to the ones found in the CEP – the outcome of the PDP process. The whole process of compiling the PDP is a joint one, where the mentor assists with target setting and action planning and relates it to the trainee's progress across the whole PGCE year.

In the second part of the PDP there are also four action plan pro formas. It is here that trainees, in conjunction with their mentors and HEI tutors, translate the targets for development into action plans to be completed within a given timescale.

On each of the pro formas there are key questions:

- What are my priorities for the next phase of my professional experience?
- How will I approach these priorities?
- What opportunities do I need to address them?
- How will I monitor my progress in relation to these priorities?

At the end of each pro forma the trainee and mentor are required to sign and date their completion and record their agreement about priorities set.

This PDP forms the backbone of the PGCE, which is a 36-course of study to take graduates with a degree relevant to teaching to the achievement of the threshold teaching competences in the standards for QTS. The standards for QTS are the basis for completion of the PDP as well as being the basis for the CEP and assist in identifying a trainee's strengths and areas for development.

There are eight points identified within the course where mentors and trainees are required to work on aspects of the PDP process together:

- Week 2: exploring the purpose and process of the PDP.
- Week 6: review progress and write an action plan for the first continuous block of teaching in the main or home school.
- Week 14: complete the second action plan in preparation for the next phase of the course where trainees are working in school and university.

- Week 17: complete the third action plan in preparation for using it as a basis for working in a second or complementary school placement.
- Week 28: back in the home school. Complete the fourth action plan as a basis for working through the final extended block of teaching.
- Week 34: review, update and complete the PDP and use this to inform completion of the CEP.
- Week 35: hand in the CEP for completion at the university and return.
- Week 36: end of the PGCE course/award of QTS.

## Commentary

The PDP is one training provider's solution to a seamless continuum leading from the beginning of the initial teacher-training programme to the point of transfer to the CEP at the award of qualified teacher status. The mentor and trainee are already accustomed to identifying strengths and areas that need further professional development well before it is time to commit pen to paper in the CEP. The areas for development tend to grow from the PDP and can be precisely framed according to the terms used in the standards. They are areas for development and some may be new in that they are applicable to the NQT rather than the pre-service. It is likely that they are aspects of teaching that the trainee has had the opportunity to work on in more than one school placement and will require refinement rather than initiation.

### How the PDP feeds into the CEP

Towards the end of the pre-service training programme the PDP is reviewed by the mentor in the main placement school. This review identifies the trainee's strengths and targets for development. A similar process has occurred at the end of each teaching block. The evidence in the PDP becomes the foundation of the practice of teaching report, which is sent to the HEI provider. The entries in the practice of teaching report relate to the standards for QTS and there are sections where the mentor and senior school (professional) tutor are encouraged to write in a more generic way about their trainee's general progress in teaching over the PGCE year. The entries in the practice of teaching report draw on wide-ranging discussions between the trainee, the subject mentor, the form tutor mentor, the senior school tutor and the HEI tutors. It is a process whereby entries are negotiated by all.

At the end of the pre-NQT year, the CEP is given to the newly qualified teachers before they move to their first teaching post. This allows the NQT to share it with the induction tutor and to reflect on its contents. In the CEP there will be between one and four statements indicating the trainee's areas of strength and up to four targets set for further development (a maximum of eight statements). The responsibility of the trainee is to 'put flesh' on these statements by considering each one and preparing for discussion and planning with the induction tutor in their school.

## The transition in mentoring

During the training period that precedes the induction year, the trainee is likely to have benefited from a substantial amount of individual personal and professional support. Some NQTs enter the profession over-confident, believing that this level of support will automatically be ongoing and some, unfortunately, will secure first teaching posts that are beyond their capabilities. The induction tutor needs to be aware of this potential mismatch in the NQT's level of aspiration and the NQT's potential to realize this. The problem is exacerbated if the culture of the school means that although help is there, it has to be sought rather than being offered. The first few months of teaching are likely to be emotionally as well as professionally demanding as the NQT adjusts to the new role. Working in an environment where help is freely offered, the NQT is much more likely to welcome advice and respond well to any support. Tickle's (1999) work is particularly helpful in setting out ways of giving support.

## Challenges and responsibilities in your mentoring

The challenges include providing not just personal but professional support. This can be difficult where the trainee is asking for assistance with only classroom management issues. You need time to get to know your trainee as a person before you start work, you need time for regular observation of classes, and time to help your NQTs build expertise in teaching their subject beyond entry level requirements for QTS. At the outset of the year NQTs may well regress to a level of proficiency that was more representative before the end of their pre-NQT year. Sensing this, your NQT may withdraw from facing new challenges with you.

Your main responsibility as the induction tutor (and we shall cover roles and responsibilities in much more detail in the next chapter) is to build professional proficiency and confidence, as well as addressing the targets within the CEP.

## What else might you need to know?

For the induction mentor to build upon the strengths of the pre-NQT training he or she needs to know what the training programme comprised. The first section of the CEP sets out this provision and would be a useful place to begin discussion with your trainee about what has gone before in the training process. Rather like a parent deciphering what lies between the lines on a child's school report, the induction tutor should probe gently into the targets identified for the NQT's continuing development. The areas of strength are probably clear from the document and the NQT may be more ready and able to identify detail, but the more problematic areas are unlikely to be so clear. A line or two of description written on a page cannot convey the full picture and needs 'unpacking'. Where the areas for development are referred to the individual statements within the standards for QTS, this task is likely to be easier for you as the induction mentor. However, it does means that you have to have a working knowledge as well as an understanding of the scope and focus of individual standards.

## The implications of induction mentoring

Any induction programme you create must be tailored to your NQT. You need support and opportunities for your own development for this role. This is a key gatekeeper role in the profession of teaching. You deserve recognition for your contribution within your school.

The strengths identified in the CEP may be highly contextualized. According to Simco (1999) (at the British Educational Research Association) the induction mentor has to assist the NQT in transferring the strengths identified to the new teaching context. The targets identified within the CEP are likely to be only partially addressing your NQT's needs in the first year of work in the teaching profession. You are reliant on the ability of the pre-NQT mentor to identify CEP targets. You needs the skills and insights to identify other targets that will enable your trainee teacher to teach successfully within the context of your school. Like your trainee, you will be a learner in the context of your mentoring.

## Getting an induction mentoring programme under way

In this chapter and the next, there are detailed illustrations of practice and analytical commentaries to help you to construct an induction programme that integrates:

- the statutory requirements for the induction year;
- the needs of your NQT within the context of your school;
- the strengths of your NQT within the context of your school;
- your own professional development.

Obviously it would be impossible within this handbook to give individual advice and you may need this. Possible sources of support and guidance for you include;

- other colleagues in your own school;
- the LEA advisor in your area with responsibility for induction;
- the helpline for any teaching union to which you belong;
- articles in the *Times Educational Supplement*;
- your local HEI provider of ITT (there may be a course for induction);
- the TTA helpline.

It will obviously be an easier matter to construct a suitable induction programme where you already know the NQT before the start of the year. Make every effort to keep in contact before and over the summer vacation. Arrange for your NQT to make a visit/series of visits to your school before the end of the summer term where you:

- discuss the CEP;
- discuss organizational aspects of teaching in your school;
- ensure your NQT has the teaching resources needed to start work;
- construct the induction programme for the coming year, at least in outline;
- clarify your expectations of the NQT and the NQT's expectations of you.

---

## Illustration

Neil has been working in school X during his pre-NQT teaching block placement. The staff have been so impressed by his teaching in class and his professional manner that during the placement he is encouraged to apply for a teaching post that arises within the department where he is training. He meets stiff competition at interview and still comes through as the best candidate for the job. After his appointment the induction process begins. Targets are set for his professional development in the weeks remaining before he reaches QTS. The subject mentor is given prior notice that he will be the induction tutor over the coming year. The subject mentor and Neil work together to identify what the targets for further development are likely to be and also how they will meet them by planning the timetable for individualized mentoring early, and integrating it into the school's induction programme for all new staff.

## Commentary

Neil and his mentor target his final weeks in school to working on precise goals; gaining valuable time and experience for the coming year. This means that his induction programme is effectively under way before September and there is a seamless transfer between the phases of his training. Neil is not the only NQT in the school and so he has the opportunity to benefit from a shared experience as well as an individualized programme. The induction programme that operates in his school has two branches. The first is the generic programme for all new staff, which includes:

- the role of administration staff;
- monitoring and assessment;
- report writing and parents' evenings;
- special needs provision;
- meeting the school governors;
- the nuts and bolts of school routines;
- summarizing the aims of the school;
- the school and the wider community.

The second is a specialized programme for all newly qualified teachers:

- planning an individual programme to support and challenge;
- using the SEN policy in practice;
- behaviour systems management;
- student improvement/target setting;
- teaching and learning styles.

During the final weeks of his teaching block, Neil is given the opportunity to work with his mentor and also with established members of staff with whom he has had limited contact up until now during his pre-service training. Next year there will be a focus group for mentors working in the school to meet up regularly and to discuss their work with one another and with a colleague from the HE with whom the school works in partnership. His mentor is well regarded in school and has further mentoring experience to draw on from colleagues within the subject department and across the school. In Neil's department there are three mentors and they will share the induction mentoring, ensuring that Neil has experience of broadening his teaching skills and building on his strengths. Next year there will be a focus group for mentors working in the school to discuss their work with one another and a colleague from HE. Some of the mentors are drawing on their mentoring for accreditation purposes at MA level.

Neil is fortunate. His school actively promotes professional development and as you can see from the illustration and commentary he is set for a successful induction.

---

Good practice in using the CEP:

- Discuss the CEP with your NQT well before the start of the school year.
- Ensure that your induction programme draws from content of his/her CEP.
- Focus mentoring on the statements in the CEP but appreciate their context.
- Delve behind the statements in the CEP by sensitive discussion with your NQT.

# Chapter 15

# Working with a newly qualified teacher

In this chapter we shall be considering the following aspects of mentoring:

- The need for the process of induction.
- TTA standards for induction.
- Setting up an induction programme.
- Identifying good practice for induction.

## The need for the process of induction

Why is a period of induction necessary? In July 1992 the Secretary of State for Education announced the abolition of the statutory period of probation for newly qualified teachers. This loss of governmental backing for the induction process has been effectively reversed and since September 1999 all newly qualified teachers have had to undergo a period of induction in schools that conforms to statutory requirements from the TTA.

In some schools NQTs have traditionally received a common training/induction programme alongside pre-service teachers. Although there are areas of commonality, NQTs have specific needs. As Tickle (1999) points out:

> We should not, I believe, simply assume that continuity is achievable in some smooth, transitional sense, regulated by the so-called standards of a Career Entry Profile. Rather we might be prepared for discontinuities; for new and radically different experiences . . .

Induction for newly qualified teachers must surely be seen as part of an endless process of continuing professional development. It is a bridge between pre-service and continuing professional education. It is part of a continuum but it is also an entity, for it has characteristics that distinguish it from other professional development. Consequently, the mentoring process is also similar and yet distinct, but how might we define the differences between mentoring within pre-service training and mentoring within the induction year? How does the mentee perceive their experiences? Although there is likely to be a feeling of collegiality within the group of trainee teachers engaged in a course of study. Rowley (1999) reminds us that 'many veterans remember their first year in the classroom as a difficult and lonely time'.

Newly qualified teachers find themselves facing very different challenges and responsibilities as they enter their first teaching post. A survey among six NQTs in one LEA revealed that one was given a temporary contract and wanted to leave – she was being harassed by the boys at her new school. A second was working in a large comprehensive school that she liked – but nobody had said they were pleased she had joined the staff and she felt unwanted. A third found herself with an OFSTED grading two points higher than her colleagues – and friction. A fourth was effectively head of German from the day she joined the school – there was nobody else to take on this role. The fifth had joined the school mid-year. She was taking all the 'sink classes' – the ones the last teacher could not cope with and the reason for her departure. The sixth had family commitments and was forced to take a temporary post to fill a maternity leave. All six in employment in one local authority area; all six facing very different pressures.

Despite their problems, these NQTs were relatively fortunate: their LEA offered comprehensive induction support. But many NQTs in other authorities continued to find themselves battling on their own. Where induction support was made available in schools, the subject content of those programmes on offer before the TTA regulations were introduced was usually focused on enabling the NQTs survival rather than promoting longer-term growth.

With the introduction of standards for QTS that call for achievement in the areas mentioned above, you might assume that problems had been solved within the pre-service training. The standards for QTS represent entry level skills and the CEP is likely to bring to the fore some of the aspects where there was sufficient achievement for entry into teaching but as yet insufficient for really effective practice in the classroom. In a sense, then, we find ourselves back at the stages of development identified by Furlong and Maynard in 1995.

However, there are distinct differences. This time the trainee has to be much more proactive about seeking assistance, perhaps taking responsibility and asking that some of the moneys allocated to school for induction be used to fund attendance at courses. In the induction year, the trainee has to meet with every class in September ready to teach, having prepared in advance and with less expectation of being spoon-fed throughout the mentoring process than previously. The emphasis is on refining skills and developing aptitudes and less on starting *ab initio*, but there will be experiences that are new to your NQT because each school context is different. Your role is to work with your NQT and to offer support in a climate of change.

Tickle's (1999) work into the induction process is a valuable starting point for understanding what are the principal differences between the pre-service and NQT phases of teacher training.

---

## Illustration

A scan of several PGCE students' Career Entry Profile entries for 1999 suggests that the induction process needs to be one of development. However, this is not to say that tutoring is just about refinement. The previous mentor cannot know the changes that your school context will require your NQT to make.

PGCE student A needs to acquire experience and expertise in areas of team management and to build on theoretical knowledge of monitoring and assessment. PGCE student B needs to develop a sense of pace in teaching, to be more demanding in terms of class behaviour, and to ensure differentiation of materials for all classes. PGCE student C needs to understand the need to break down skills into constituent parts, to develop a wider range of teaching styles – especially problem solving – to develop use of voice, tone and projection during the lesson.

Career Entry Profile entries stress the need for experience, for development, for practice. Although the 63 standards for QTS may have been achieved there is much for you, the mentor, to do during the induction year if your trainee teachers are to have a sound basis for their teaching and avoid low motivation and energy and early burnout. So much for the measurable aspects of teaching against competence statements, but what aspects of teaching may not have been addressed at all or addressed insufficiently within pre-service training? The effectiveness of training courses should be more consistent given the standards for QTS but it is still likely that it will vary enormously from trainee to trainee and from provider to provider.

## Illustration

> . . . there is no effort made in some PGCEs to introduce you to pastoral things such as form tutoring. Another hazard for beginners is parents' evening. Good courses train for it.
> (*Times Educational Supplement*, 10 January 1997)

> I wasn't prepared for the drain of stamina – this is much more tiring (than the PGCE year) . . . the level of support is good in school which is good – I need it. I feel very vulnerable . . .
> (NQT, 1998)

### Commentary

Clearly just knowing that an NQT has achieved QTS through gaining a postgraduate qualification cannot be taken as a guarantee that all NQTs have had a similar content or approach within the programme of pre-service training. Mentoring requires an audit of all NQTs' needs and a clear indication of what requires development before they can undertake effective teaching.

---

## Differences in responsibility

A change of culture results from moving into the 'real' from the 'practice' world of teaching and is one of the most striking aspects of the transition from pre-service to newly qualified status. Mistakes will have almost certainly been made in the placement schools, but these could be left behind and attributed to experience. In the first teaching placement newly qualified teachers are faced with a pressure of greater accountability for their actions, and this time they are staying on.

## Why do NQTs need standards for induction?

If the links between initial teacher education and the Framework for Professional Development introduced in 1998 are to be improved there has to be continuity of practice. Thus the standards for QTS form the basis of the CEP, which in turn forms the basis of the induction year – so the standards for QTS and additional induction standards comprise:

- Planning, teaching and class management.
- Monitoring, assessment, recording, reporting and accountability.
- Other professional requirements.

## Mentors need professional development too!

The continuing professional development of newly qualified teachers is only half of the training equation. DeBolt (1992) highlights also the need for training and preparing induction mentors. If you have been involved in pre-service training there is a lesser challenge in undertaking mentoring within induction but there are still substantial differences, which may include the absence of a well-structured programme and help on hand from outside school if necessary!

You are more likely to need professional development if you are unfamiliar with the trainee's pre-service training course or where you have not had the opportunity to mentor previously for ITT. In particular you may find the statements of competence for NQTs a daunting prospect!

## The induction of NQTs – what the TTA does specify

Since June 1998, all UK providers of ITT have had to provide NQTs with a TTA CEP when they successfully complete their ITT. Only those who gain QTS receive a CEP. Although we addressed the use and purpose of the CEP earlier, it is worth restating that it is designed to support the transition from ITT to teaching and continuing professional development, by providing information, in relation to the standards for the award of QTS, about a new teacher's strengths and priorities for development; and the CEP requires a new teacher to set objectives for that development, as well as an action plan for induction.

Under the statutory induction arrangements, the induction mentor is responsible for supporting the NQT and helping to implement a programme of monitoring, support and review based on the action plan set out in the CEP. All those who complete a course of ITT satisfactorily and are awarded QTS have been assessed against the standards for the award of QTS and judged to have met them all successfully. The main headings used in the standards for the award of Qualified Teacher Status are set out below:

- knowledge and understanding;
- planning, teaching and class management;
- monitoring, assessment, recording, reporting and accountability;
- other professional requirements.

The TTA sets out the induction tutor's duties stage by stage during the induction period. What is particularly useful in the TTA's support materials for the induction period is the pack that accompanies the overview of the legislation governing the induction process. We examine the way the TTA has set out the responsibilities of the induction tutor below. After that we look at aspects of the responsibilities the TTA does not mention. Legislation cannot cover everything!

As well as any tasks delegated by the headteacher, the responsibilities of the induction tutor include:

- Making sure the NQT knows and understands the roles and responsibilities of those involved in induction, including his or her own rights and responsibility to take an active role in his or her own professional development.
- Organizing and implementing, in consultation with the NQT, a tailored programme of monitoring, support and assessment (see Circular, paragraphs 40–54), which takes the action plan set out in the NQT's CEP forward in a flexible way and that takes into account the needs and strengths identified in the profile, the induction standards, and the specific context of the school.
- Coordinating or carrying out observations of the NQT's teaching and organizing follow-up discussions with the NQT.
- Reviewing the NQT's progress against objectives and induction standards.
- Making sure that the NQT is fully informed about the nature and purpose of assessment in the induction period.
- Ensuring that dated records are kept of monitoring, support and formative and summative assessment activities undertaken, and their outcomes.

As well as providing formative assessment, the induction tutor will, in many cases, be involved in the formal, summative assessment at the end of the induction period. The support and assessment functions may be split between two or more teachers if this suits the structures and systems already in place in the school. However, if a school takes this approach, one person must be identified as having day-to-day responsibility for coordinating the induction programme, and it is important for responsibilities to be clearly specified at the outset. In a large school, a senior member of staff may take an overview of the induction arrangements for all NQTs across the school. This would be in addition to the induction tutors, who have direct responsibility for tailoring the induction programme of individual NQTs.

This is a pretty comprehensive overview of your role but as a mentor you should also be aware that your NQTs will develop through a series of jerky stages and not in linear progression! You need to plan for a shifting in the mentoring relationship between you and your NQTs. At first they will often be dependent but they need to move away over time!

Put yourself in the shoes of your NQTs – excited and daunted, full of anticipation and possibly not a little dread. Think about how your role looks to them.

## What NQTs need to know about – and the TTA doesn't specify

- The pupils – which classes they are taking; who taught them last year, and so forth.
- Stock availability and management; stationery; resources and equipment.
- Reprographics; staff computing facilities; support staff.
- Keys, lockers, personal work space, car parking.
- Organization of the school day; registration and tutorial duties; teaching groups; staff duties at break, lunch, after school.
- Teaching timetable, mark book, routines and teaching room(s).
- School, faculty and departmental policies on: the curriculum – syllabuses, schemes of work; homework; assessment and record keeping; equal opportunities; special educational needs; pastoral work; discipline; bullying, and so forth.
- Acronyms commonly in use – SMT, SoW, SDP, TGIF!
- How to manage stress and time wisely.
- Extra-curricular activities.
- The management structure of the school; key personnel; line managers, and the support that can expect; names of departmental colleagues and others taking the same year group; union representatives.
- The school brochure, the staff handbook; LEA handbook – LEA policies, curriculum support, professional development, support for NQTs, and so forth.
- Non-teaching staff – clerical support, science technicians, library and reprographics personnel.
- Details of staff social activities!

(Adapted from Capel, Leisk and Turner, 1995.)

Good practice in induction mentoring means constructing a programme that is at once comprehensive, challenging, supportive and sufficiently flexible to meet individual needs.

---

### Illustration

Here is an overview of the induction programme for one large secondary school:

Newly qualified teachers' induction programme 1999–2000

All meetings are on Wednesday. For NQTs, these meetings should take priority over other commitments, including sports fixtures and musical events, unless previously negotiated with the induction tutor.

Meetings will start at 3.45 pm with a progress review with NQTs followed by the formal session between 4.00 and 5.00 pm.

Autumn term

8 September – introduction to the ICT network with the x systems manager.

15 September – marking and assessment with X (deputy head).

29 September – the pastoral system with X (senior teacher).

13 October – progress review tutorials with X.

3 November – managing behaviour with X (senior teacher).

1 December – summative assessment tutorial with X.

15 December – review

Spring term

2 February – summative assessment tutorial with X,Y and Z and head of faculty.

29 March – progress review meeting with KB. Introduction of research task.

Summer term.

17 May – progress review meeting with X. Discussion of research task.

20 June – new staff induction 2000. Presentation of research task.

28 June – final assessment tutorial with X and Y.

## Commentary

There are a number of key features of good practice in this illustration:

- The meetings should take priority over other commitments in school.
- The first term is quite tightly structured.
- There is scope in the second term for a programme tailored to meet the NQTs' individual needs in terms of personal and professional development.
- There is a flexibility as well as a clear organizational framework.
- Although replaced by 'X, Y and Z' here, key members of SMT are involved in the programme.
- The content of the programme is varied and broad.
- There are regular tutorials with the senior member of staff in addition to those with a subject departmental mentor.
- There is an emphasis on continuity and progression – a good sign!

This school has been involved in providing PGCE placements for trainee teachers for a number of years. It has developed the induction programme on the basis of this mentoring, which has been conducted in partnership with several HEI providers. The professional tutor for the PGCE trainees is the creator of this induction programme and reflects wide expertise in mentoring.

The illustration above is just one example of how to approach the construction of a suitable induction programme for newly qualified teachers.

---

Central to any programme must be the recognition that:

- All newly qualified teachers are unique.
- No one mentor can solve all of an NQT's difficulties.
- Mentors need support – from peers, LEA advisors and senior colleagues.
- Mentors need to be reflective practitioners if they are to inspire their NQTs.
- All NQTs should come to realize that meeting competence statements is only the beginning! (Teaching is 'an open complex form of skill' – Tomlinson, 1995.)
- Mentoring must be adaptive – all NQTs already possess transferable skills.
- Even a structured mentoring programme may not address all of the NQTs' needs.
- The induction programme set out by the TTA is a minimum entitlement for NQTs.

Questions to ask yourself:

- What are my NQT's strengths identified in the CEP?
- What are the priorities for development given in my NQT's CEP?
- What opportunities will I have to enable my trainee to develop in these areas?
- What is likely to be unfamiliar to my NQT and what might be problematic?
- How can I ease my NQT's journey through a very tough year?

Good practice for induction:

- Mentoring is similar but not the same as that for pre-service teachers.
- Induction programmes must draw from the NQT's needs in the CEP.
- The TTA has set out the statutory requirements for an induction programme.
- An induction programme should go beyond the TTA's statutory requirements.

# Chapter 16

# Appraising competence

> The new pay system for all teachers comes into effect from 1 September 2000. Its main features include: the performance threshold for experienced classroom teachers, giving access to an immediate salary increase of £2000 and, in future, further pay points on the new upper pay range for substantial and sustained achievement. (www.dfee.gov.uk)
>
> 'All appraisal procedures aim at developing people . . . by using information about the behaviour of people at work.'
> (Randell, Packard and Slater, 1984: 12)
>
> In this chapter we shall be considering the following aspects of mentoring:
>
> - The role of appraisal in professional development.
> - Stages in the appraisal process.
> - Linking the TTA standards for QTS to appraisal.
> - Identifying good practice in appraisal.

## The dual role as appraiser and assessor

Appraisal in ITE should be developmental and can be defined as the evaluation of professional performance – in this case the performance of 'being a teacher' (rather than simply the act of teaching) of the trainee. This is carried out by the mentor with the trainee. In the context of TTA standards it has come to mean:

- Recognizing the achievement of the trainee against competency statements.
- Identifying the potential of the trainee for further development within specific competency statements.
- Helping the trainee to see ways of overcoming difficulties in meeting some of the statements of competence.
- Setting appropriate targets for moving the trainee to a higher level of competence or a new area of competence.
- Alerting those who can assist trainees to develop or extend an area of competence their need for assistance.
- Assisting trainees in turning their targets for professional development into action plans and enabling the action plans to work.

Appraisal, like assessment, is part of every teacher's role, but how do they differ? The answer lies in the term 'appraisal', which derives from the archaic form *apprize*, by assimilation to praise. Appraisal is about valuing; assessment is about estimating quality. For the mentor working in school, appraisal is an essential aspect for moving the novice towards professional proficiency. Appraisal is sometimes confused with assessment but the goal of assessment is to measure rather than to improve. As a school-based mentor you are an assessor as well as an appraiser – a gatekeeper into teaching as well as being a personal enabler or change agent. As a gatekeeper, you have a responsibility to share the criteria by which your trainee will be assessed in a formative way as well as in a summative way. There is likely to be assessment at regular intervals, perhaps through specified course work in terms of assignments or school-based tasks, but this is in addition to the appraisal of teaching.

The developmental nature of appraisal is apparent in the Education (School Teacher Appraisal) Regulations 1991 (HMSO), which aim to:

- recognize the achievements of schoolteachers and help them to identify ways of improving their skills and performance;
- identify the potential of teachers for career development, with the aim of helping them, where possible, through appropriate in-service training;
- help schoolteachers who are having difficulties with their performance through appropriate guidance, counselling and training.

Although there are advocates, appraisal has been a rather contentious issue for a long time in the history of teaching in England, especially where it has been linked to performance-related pay, which is the current situation in England. Demands for a system of schoolteacher appraisal gathered momentum in the 1980s with the belief that appraisal was necessary to improve the quality of education. This report advocated 'systematic performance appraisal, designed to bring about a better relationship between pay, responsibilities and performance, especially teaching performance in the classroom' (DES, 1985: 55–56).

Certainly appraisal is a complex issue and its very complexity has added to the difficulties that are sometimes associated with it. However, if the summative assessment of trainees at the end of training can be problematic for some mentors, the appraisal of their progress over the period of the ITE course seems to be less so.

## Getting the balance right

Appraisal within mentoring should starts out with a process of identifying trainees' potential for development with the aim of helping them, where possible, through setting targets and facilitating their achievement. Appraisal means working alongside. There is an inevitable shift in relationship from working alongside and taking on a position of authority to assess. If the personal relationship between a trainee and a mentor has become too close, too friendly or too relaxed, the shift to assessment can cause a fracture of trust when the assessment reports have to be completed by the mentor. Friction results if the trainee does not realize the dual role of the mentor as appraiser and assessor. This duality must be clear from the outset.

## How appraisal affects your work as a mentor

Brooks and Sikes (1997) set out three main categories of mentoring:

- The apprenticeship model with the mentor as skilled craftsperson.
- The competence-based model with the mentor as trainer.
- The reflective practitioner model with the mentor as co-enquirer.

In practice, the distinction between the three is not quite as clear cut as this definition might suggest. Engaging in good practice in mentoring will involve embodying all of these models. There will be occasions when you will demonstrate something in your mentoring for your trainee to copy to as high a professional standard as possible – filling in the class register is a simple example. Where you are taking your lead from the standards for QTS as a basis for appraisal, you will be working for much of the time in your mentoring capacity as a trainer, making sure that your trainee develops skills set out in the various statements of competence. Planning a lesson is a case in point. The way in which a lesson is planned varies from teacher to teacher – there is no easily definable way of doing it – hence the different slant from the apprenticeship model, which is built on imitation. The origin of the reflective practitioner model lies in the work of Donald Schön (1987) who identified two types of reflection within professional artistry. He distinguished between reflection-on-action (reflecting on practice after the event) and reflection-in-action (reflecting on practice during the event) and although he was not working with classroom teachers when he developed his model it has been widely applied to training novices learning to teach in schools.

How do these three interrelated models affect the appraisal undertaken within mentoring? What difference does it make if there is just one form of mentoring or three? It means that you can develop a wider range of skills for appraisal because you see more than one dimension to mentoring. No matter which model you begin with – and this will depend largely on the previous teaching experience, or lack of it, that your trainee brings to ITE – the first aspect of appraisal is to carry out an audit. The questions in the audit will vary according to whether you appraise in terms of apprenticeship or competence-based teaching, and the degree of reflective practice.

### *Appraising your trainee as apprentice*

Points to consider in your audit:

- Does your trainee have the capacity both mental and physical to carry out set procedures in school?
- Does your trainee understand how to carry out specific procedures in school, for example how to complete a report form to send to parents?
- Does your trainee know how to carry out procedures but needs additional training to do them to a higher standard?

### *Appraising your trainee against competence statements*

Points to consider in your audit:

- Is your trainee aware of the competence statements against which appraisal is taking place?
- Has your trainee demonstrated the potential to meet the levels of competence set out in the predetermined criteria?
- How many of the targets that the trainee has been working towards within the statements of competence are being met?

### *Appraising your trainee as a reflective practitioner*

Points to consider in your audit:

- Does your trainee understand how reflection affects practice through considering what has been done as a basis for improving future actions?

- Does your trainee understand the difference between reflection-in-action (thinking on one's feet to improve what is occurring) and reflection-on-action (thinking over what has occurred and how it could have been improved) – and using both as a basis for future improvement?
- Does your trainee engage in reflection and if so how could this reflection be used to better effect in classroom teaching?

A good mentor is able to work in all three modes as we can see in the following illustration.

---

## Illustration

P has two mentors – one is in effect a trainer within her subject department, working closely with her to secure achievement against the standards for QTS. This mentor also works as a role model – teaching P specific techniques in aspects of classroom teaching where apprenticeship is appropriate; and assessing pupils' work in examinations. Her other mentor works with P to encourage her to become more thoughtful about how she approaches her work – he is a reflective mentor – not concerned with the day-to-day practicalities in a subject department but working with P so that she will develop an understanding about education. This is essential if she is to balance the other aspects of mentoring occurring in her subject department where she is being trained to teach.

## Commentary

Within the hustle and bustle of the subject department there are few opportunities to enable time for lengthy reflection on teaching. The mentor is a reflective practitioner herself – but has little opportunity to discuss how she reflects in and on her own practice. She makes a point of using a reflective approach in offering feedback on lesson observed but knows that making explicit how she reflects is not her strength. In effect she has devolved this responsibility to a second mentor who can stand back from the immediacy of the classroom situation. He and the subject mentor liaise to ensure that P has a balanced mentoring programme and both mentors share the work of appraising P. Working as a threesome, they negotiate to examine her strengths and the areas she needs to develop, and they report back fully on the outcomes to one another.

---

If teaching was learnt by simply imitating what an expert teacher does it would be easier to perfect and to appraise. Teaching cannot be defined in purely behaviourist terms but nevertheless there are aspects of it that are mechanistic. Managing the paperwork relating to recording pupils' grades might be seen in this light. There are aspects of time management that trainees can be taught and that will assist them to work efficiently – allowing time for packing up at the end of a lesson, and checking homework diaries quickly. Teaching is a highly complex art – but some aspects of it are not complex; they are repetitive and open to being learnt by imitation of a model.

The problem with adhering to a single model of mentoring – competency, apprenticeship or reflective practice – is that this necessarily excludes a raft of skills that are essential to good practice within the profession of teaching. The reason for appraisal across a range of approaches is rooted in the fact that teaching itself is neither just apprenticeship nor meeting competences, and reflection too is well served by having a framework of targeted questions through which to stimulate thought. You will need to be flexible in your approach to appraisal to make the process worthwhile.

How will you appraise your trainee's ability to learn to teach? You need to:

- appreciate that there are different ways of learning;
- appreciate how your trainee engages in learning;
- mentor in a way that enhances your trainee's learning;
- mentor so that your trainee develops mastery in artistry.

Tomlinson (1995) distinguishes between several types of learning in teaching:

- Learning to distinguish and unravel the many embedded elements and processes that tend to be involved in any aspect of teaching.
- Learning to combine and integrate these elements.
- Learning awareness and learning how to read and monitor teaching situations.
- Learning what to do in them – learning how to act, anticipate and respond.

It is this final category of learning that relates most closely to the apprenticeship and competence models of initial teacher training and in a sense this final aspect is the easiest to appraise, but appraisal must take account of all of these aspects of learning and these aspects will be apparent at different stages in the training process.

## Stages in the appraisal process

Appraisal of your trainee has to be approached in a series of steps that match his or her development as a teacher. It takes place where your trainee has learnt:

- to distinguish and unravel elements and processes in teaching;
- to combine and integrate these elements;
- to read and monitor teaching situations;
- what to do in them, learning how to act, anticipate and respond.

There are stages in appraisal just as there are stages in mentoring that correspond to stages in trainee–teacher development. Bearing in mind these stages of learning, the mentoring you undertake must be timed at appropriate intervals across the ITE course. If the appraisal sessions are too regular the trainee will not have an opportunity to put action plans into effect. If appraisal sessions are too infrequent momentum will wane. It is important to encourage trainees to appraise their own progress as well as to appraise as a more objective observer. Keep a professional development diary that looks back to what has been achieved, that looks at the present to consider if progress is satisfactory, and that looks to the future to set goals and to frame these in terms of action plans.

In the following example we shall consider aspects of the professional development plan for the PGCE at the University of Bath and how it is built on a series of reviews.

---

### Illustration

The 36-week PGCE course is arranged around serial blocks (where there are university teaching days and school-based teaching days in the same week) and teaching blocks (where the novice teacher spends all of the weeks in the school). The rubric in the professional development plan states that:

> A good way to use the document is to review periodically the Standards (perhaps twice during each Serial Phase and every fortnight during teaching blocks) by going through the groups noting for each Standard what progress you have made in relation to achieving it and citing source(s) of evidence to support your claims.
>
> After periodic reviews of your progress it will be appropriate for you to write an action pan with your mentor and tutors to support

your development in the next phase of the course. This will be appropriate in the transitions between each phase or transition in the course.

Starting points will vary according to individuals' needs, school contexts and subjects.

### Commentary

What stands out about the process is the stages through which it passes and the framework for professional review and development. Reviews are made regularly by the trainee alone and also by the trainee in conjunction with mentor and HEI tutors. In this way there is a process of self-appraisal, of valuing what has been achieved and of setting realistic targets for development and using the standards for QTS as a basis.

## Linking the TTA standards for QTS to appraisal

The standards for qualified teacher status provided by the TTA are for exit-level assessment. In many ITE courses, these have become the basis for appraisal of the trainee during a training course. If handled sensitively these statements can be used effectively in appraisal but it is important to recognize that this was not their intended purpose. Appraisal against the TTA standards means praising what trainees can do rather than berating or marking them down for what they cannot. It is a model of positive and constructive mentoring through looking for ways to develop the trainee's skills. Difficulty can arise for the mentor if the trainee resists or does not understand that the standards are a framework for learning and development. This is compounded when summative assessment is not overtly linked to the statements of competence, causing the trainee to lose confidence because monitoring against appropriate targets has not occurred. It is vital for the mentor to realize that not all trainees have the potential to achieve all of the competences in the standards for QTS and that this is not the mentor's failing.

With regard to primary courses:

> The initial teacher training curricula specify the essential core of knowledge, understanding and skills which all trainees on all courses of initial teacher training must be taught and be able to use in relation to English, mathematics and science. The training curriculum for the use of

> information and communications technology in subject teaching specifies the essential core of knowledge, understanding and skills which all primary teachers on all courses of initial teacher training must be taught and be able to use . . .

For secondary courses:

> The initial teacher training curricula specify the essential core of knowledge, understanding and skills if they are specialising in English, mathematics or science. The initial teacher training curriculum for the use of information and communications technology in subject teaching specifies the essential core of knowledge, understanding and skills which all trainees on all courses of initial teacher training must be taught and be able to use . . .

The rubric in Circular 4/98 is carefully worded to prompt teacher trainers to realize that the competence statements within the ITE curricula are not the sum total of what a trainee needs to know through training but rather that providers should use the curricula as a basis for designing courses that are coherent, intellectually stimulating and professionally challenging. Appraisal against these criteria will necessarily involve looking beyond what the trainee does to what the trainee thinks and how this thought can be translated into improving professional development by reflection.

In order to enable trainees to appreciate the need to reflect as well as to imitate and to develop practical skills, there is something to be said for delaying their exposure to all of the of competence statements against which they might be appraised and will eventually be assessed. The danger of setting down all the standards as a basis for development from the outset of a training course is that trainees will simply feel completely overwhelmed and will just panic – the reaction it is imperative to avoid.

---

## Illustration

Trainees are encouraged to compile a portfolio that demonstrates their professional development over the training course. They are given a copy of the standards against which they will eventually be assessed but the mentoring and appraisal within the mentoring is geared to empowering – showing how the portfolio can be used to demonstrate proficiency rather than using individual lessons taught by a trainee as a basis for a tick-list approach to competence.

## Commentary

The portfolio approach (this one from a SCITT consortium) enables the trainee and the mentor to assess progress over time and over a range of experiences that they decide are appropriate. Within the portfolio there are likely to be videotapes of classroom teaching, samples of teacher work (including reflections on the effectiveness of teaching strategies) as well as samples of students' work. It is the intelligent application of such knowledge that is the essence of good practice.

---

Good practice in appraisal:

- Start with an audit of your trainee's strengths and areas for development.
- Agree targets for development and action to be taken in a specified time.
- Monitor progress against the targets set on a regular basis.
- Use the strengths and areas for development identified to lead into assessment.

# Part 5. DEVELOPING COMPETENCE THROUGH MENTORING

# Chapter 17

# Knowledge and understanding

In this chapter we shall consider the following aspects of mentoring:

- Knowledge and understanding in the TTA standards for QTS.
- Knowledge and understanding as a basis for selection of trainees.
- Knowledge and understanding developed during the ITE course.
- Identifying good practice in developing knowledge and understanding.

## Knowledge and understanding within the TTA standards for QTS

The current TTA standards for qualified teacher status require pre-service teachers to demonstrate that they have sufficient knowledge and understanding to enable them to teach confidently. For those training to work in the primary sector, this means the core and specialist subjects in their training. For secondary sector trainees this relates to their specialist subject at a standard equivalent to degree level. The requirements that are currently set out in Circular 4/98 are under review and will be replaced by an amended Circular to be in force by 2002. It seems inconceivable that such a pivotal aspect of training will disappear in revisions.

## Selecting candidates for ITE

Before any applicants arrive for interview for a place on a teacher training course they will already have been through a selection process. The Graduate Teacher Training Registry or GTTR sends detailed application forms completed by each potential trainee to the training institution. This may be the first occasion upon which most trainees have been able to demonstrate that they have sufficient potential in terms of knowledge, aptitude and skills to enable them to teach. The GTTR form requires applicants to supply details of their qualifications in relevant subject areas. For secondary phase applicants, this means they must record whether they have passed examinations in mathematics and English to GCSE/O level or equivalent and that they have a first degree in a subject relevant to teaching. Candidates who do not have maths at an appropriate level may be entitled to sit an examination at the HEI that will confirm their proficiency. For middle and primary school applicants the GTTR form requires that applicants demonstrate they have attained qualifications in mathematics, English and science and hold a first degree. All candidates must supply a confidential reference, which should give a positive indication of their personal and professional suitability to become a teacher. This reference is usually from an academic source. They have to give details of a second referee as well. On the basis of the information given in the GTTR form candidates are selected for interview.

This is the start of a process that filters applicants who show they have sufficient knowledge and understanding to become potential teachers. Other details taken into consideration when shortlisting are:

- Evidence that a candidate has some understanding of working with children.
- Experience of teaching with children or adults.
- Commitment to working with children.
- Commitment to teaching – evidence of wanting to teach.

Some candidates write with passion and some in a more detached style. Some express their delight in sharing their subject knowledge and some say that they want specifically to work with children so that they can share the love of their subject in the classroom. Occasionally, some applicants make scant reference to either children or subject knowledge in their application forms and they rarely reach the interview stage!

At the interview, where schools are working in partnership with HEIs, there will often be representatives from both sides of the partnership – a mentor from a school and a higher education tutor. Where the school alone provides the training the course co-ordinator and a subject representative or classroom teacher usually attend. In both scenarios the interviewers are intent on discovering from the outset whether the potential entrant to the teaching profession has sufficient knowledge and understanding to become a teacher and, equally importantly, the personality to complement these. Sometimes there are disappointments. Hopefully the strength of the details given by applicants on the GTTR form will be borne out by their performance at interview. However, there are occasions when a candidate who appears to make a strong application on the basis of qualifications fails to make a convincing case at interview. Where this occurs it is essential to provide full feedback to the candidate.

As we shall see in this first illustration, some candidates certainly have the knowledge and perhaps the understanding to become teachers but they fail because they overlook the commitment that they must convey if they are to be accepted for an ITT course.

---

## Illustration

P has been following a first degree course at the same university where he is now applying to become a student on the PGCE course. He is likely to be awarded at least an upper second in his final examinations. His degree will enable him to teach to sixth-form level in a subject directly relevant to his chosen career and the reference assures the selector that he has the personality, health and persistence to teach. The interview is in two parts – first there is a general introduction to the course, where all the candidates assemble to learn about what their training will involve, before a second part where they have individual subject interviews.

P sits quietly and thoughtfully through the first part. He asks questions relating to the timing of school day and seems concerned about what time teachers can leave school. As he is walking with the tutor to the second part of the interview she chats about his application. 'Why have you applied to this PGCE course?' There is no hesitation before the reply: 'Because it will allow me to continue with my dancing – I am captain of the Latin American Team for our students and I want to be in this town.' Hardly an auspicious start. In the course of the interview the two interviewers – a university tutor and a school-based mentor – repeatedly broach the subject of what is bringing this candidate to apply for this PGCE. 'Are there examinations?' 'No – continuous assessment.' 'Oh that's good – because I have competitions during the year for my dancing which I cannot miss . . .!'

## Commentary

Needless to say the candidate was unsuccessful at this interview. He contacted the tutor who had interviewed him in astonishment upon reading the letter telling him that he had been unsuccessful in gaining a place on the following year's PGCE course and asked for a debriefing session. At this follow-up session the reasons for his rejection were clearly stated – with advice on how to fare better at the next interview:

> Remember that you are on interview all day – not just at certain times. Prepare for an interview by looking at the course documentation and be ready to give clear reasons in professional terms about your choice of the particular course rather than another. During the interview, work from the strengths in your professional base and above all – sort out your career priorities before considering taking on a very demanding career. Why not reapply for next year's course? Keep in touch this year.

There was a long pause . . .

> Thank you for telling me. I hadn't looked at it that way. I'm shocked by not being taken on the course – it's made me even more determined to be a teacher now. I've been thinking about it. I've decided to take a year out – go and do some work in a school and re-apply next year.

Although it may seem strange at first sight that such an able candidate should be so naive at interview, it is vital to look at the context here. This student is not ready to enter teaching but may well have the potential to develop into a teacher. The course is too short to take the risk that he will do so and the obvious solution is to enable him to choose to delay making an application. Will he come back? Yes, he probably will and it is worth investing time in debriefing and giving advice to keep this opportunity available to him. He needs to be in the locality for the foreseeable future and time spent in schools will help him to decide exactly what career he wants.

## Getting involved in selecting future teachers

How can mentors assist in a situation like this? One of the most important ways of enabling potential applicants to the teaching profession is to become involved in organizing 'teacher taster days'. These are funded by the Teacher Training Agency and usually attract considerable interest when advertised in the local press and through school networks. The sessions include an introductory talk – what does it mean to be a teacher in today's schools? There is an opportunity to spend time in schools talking to mentors and senior staff and to observe lessons to acquire a feel of how it is to teach there. Mentors have a pivotal role to play – recent media reports would deter even the most stalwart applicants from joining the teaching profession!

Mentors should look for opportunities to run interviews and assist with them. By shadowing an experienced interviewer mentors can develop suitable knowledge and understanding to enable them to take over interviewing. If you are not approached to assist in interviewing it is worth considering a more proactive approach than waiting to be asked. Contact the main interviewer and ask if you can assist.

So what is the interviewer looking for in terms of subject knowledge and understanding at the threshold of entry to teaching?

When potential applicants to ITE courses contact the ITE provider they need clear unequivocal advice. How to apply? Where to apply? They also need information about the expectations the provider has in terms of qualifications for suitable applicants. This will minimize the time wasted by applicants who do no meet basic entry requirements and by busy ITE interviewers.

---

## Illustration

The following is the documentation sent to potential applicants for a middle year's PGCE course at the University of Bath in the 1999–2000 year. Similar subject-specific material is sent out to all PGCE applicants where appropriate:

> There is an expectation that you should have studied sufficient mathematics at A level and degree level, in order to feel confident in your own expertise particularly in relation to the topics covered in the core syllabus for A level mathematics. We recognize that many good candidates will not have a degree in mathematics explicitly but will have gained their expertise in a related discipline. Whatever your background we should like you to carry out an audit of where you see your strengths in mathematics and areas which you may need to develop in relation to the core syllabus and that of the National Curriculum – please bring audit(s) completed to interview.
>
> If you are invited to assist at interview what can you expect to be asked to do? It is important for you to find out before you arrive

and to have a good look at copies of the GTTR forms for the candidates that you will be interviewing in advance of the session. The HEI or SCITT may provide you with guide questions like the ones given below. If not, these will provide you with some ideas about questions to ask.

---

## Questions at interview relating to knowledge and understanding

Questions at interview should address these areas: subject; philosophy: ideas about teaching: relevant prior experience; school experience.

### *Subject questions*

- Degree content.
- Subject audit and match with National Curriculum.
- Discussing any remedial action needed.

**Examples of questions:**

- Tell me about your (subject) background. What has been the most interesting thing that you have studied?
- Tell me about your dissertation. Why did you choose this topic? How does this relate to the nature of (subject)/your view of (subject)?
- Which units of your degree course did you particularly enjoy? What were the key issues?
- Has your (subject) background provided you with a good foundation for teaching (subject)? Which areas would you find most difficult?
- What areas of your subject do you anticipate you will need to work on?

### *Philosophy of teaching questions*

- Motives for teaching.
- Purpose of the subject in school.
- Why teach the subject?

**Examples of general questions:**

- When did you first think about teaching as a career?
- Why do you want to teach?

- What personal qualities do you have which will help you in teaching?
- What is the purpose of teaching (subject) in schools?
- How would you defend the place of (subject) in the curriculum?

**Examples of questions (more specific, tackling potentially controversial areas)**

- What is your opinion of the role of X in (subject) teaching?
- What are the problems in teaching X, and how would you approach them?

### *Ideas about teaching*

- Initial ideas for activities and resources on a favourite topic.
- Classroom management situations.
- Situations beyond the subject classroom, including tutoring.

**Examples of questions**

- What would be key learning outcomes if you taught X to 12-year-olds?
- What resources/activities would you set up for the children to meet these learning outcomes?
- How would you make the work exciting and relevant to the pupils?
- How might you approach the teaching of specific content/skill/process in your subject area?
- How might you deal with a difficult classroom situation (for example wide ability range in the class; discipline and classroom management)?
- How do you view the role of a tutor? How might you deal with a difficult tutoring situation (for example a racist comment or bullying)?

### *Relevant prior experience*

- Jobs/other experiences relevant to teaching.
- Change of direction.

**Examples of questions**

- What skills, relevant to teaching, have you developed in your previous work?
- What experience have you had (working) with young people?
- Why did you decide to change direction?

### *School experience*

- Own experience of teachers and learning at school.
- Impressions of recent observation in school.
- Comparing schools then with recent observations.

**Examples of questions**

- Have you been taught by any particularly good (or bad) teachers? What made them good (or bad) teachers?
- Describe your best/worst learning experience. Analyse the reasons for this being the case.
- What pupil learning was being targeted in the lesson you observed?
- How was the lesson structured?
- How might you do it differently?
- How does the teaching of your subject now compare with when you were at school?
- How do your observations differ from your experiences as a pupil?

## Knowledge and understanding developed during the ITT course

At the outset of a course of initial teacher training be it by graduate training route, Bachelor of Education (BEd), Bachelor of Arts (BA) or Bachelor of Science (BSc) with QTS, or PGCE, it is essential to instigate not only a subject audit but also an audit of the trainees' awareness of where they are starting from – their understanding as prospective teachers. The training will be largely individualized to their needs during mentoring and you need to know about the trainees' subject knowledge and their understanding of what it means to teach their subject in a school, and to understand how they have been taught and have learnt a particular subject in the school curriculum. One way of finding out about knowledge and understanding is to ask targeted questions but there may be limited time in your mentoring to get an in-depth view. Another way is to set an assignment or to ask if you can mark a relevant assignment set by the HEI tutor.

In the next illustration, again taken from Bath University, we can see how an assignment set early in the ITE course can help you to find out more about your trainee's subject knowledge and understanding.

## Illustration: PGCE in (subject) – preparatory assignment

'Where I am starting from as a (subject) teacher?'

This is the first course assignment and enables you to describe your current ideas and thoughts on [subject] in preparation for the work we shall carry out on the rest of the course. The intention is that you try to understand and bring to the surface your own ideas and preconceptions you might have about teaching [subject] and which you are bringing to the course. Such ideas are often tacit and difficult to articulate but are nevertheless influential in determining how and what you learn from the course.

We are sending you this in advance of the start of the course as it will require you to spend some time clarifying your own thoughts and we shall also be asking you to use it as the basis for some input in one of the early mathematics education sessions. Your writing will form the first part of your [subject] portfolio. There is not a rigid word limit but as a guide you should aim for two to three sides of A4. Do not get anxious about this though – there are no right and wrong answers and the aim of this assignment is to get you make explicit and begin to explore the views you already hold on [subject] and the teaching of [subject] as a result of your own experience.

You should describe your personal philosophy and beliefs about [subject] teaching and learning. Structuring under the following headings will help you to do this – use the questions as a guide as to what to include rather than a list of questions.

*Myself*

Describe your own [subject] background and include your school and higher education experiences. What do you feel about (subject) and why do you think you feel this? Have you always felt like this? What was your experience of school like? What were your teachers like? How were you taught? How did/do you feel about this? Try and think of a particular incident from your experience that stands out to you as being significant and explain why.

*Pupils*

How do pupils learn [subject]? Why do pupils often not like [subject]? Why do some pupils find [subject] easy and some find it difficult?

*Teaching*

Why do we teach [subject]? What is the purpose of learning [subject]? What sorts of things ought teachers to be doing when they are teaching [subject]? How do we teach [subject]? How, for example, might you attempt to teach [difficult aspect of subject]?

*Reflecting on writing*

Comment on this assignment. How did you feel about being asked to do it? How did you tackle it? Have you learn anything by doing it? Why were you asked to do it?

## Commentary

The 'good practice' embodied in the details of the assignment is borne out by the number and focus of the questions posed to the trainee teachers, which force them to consider not only their own experience and how this is likely to colour their expectations of themselves and their work, but also why teaching their subject should be included in the curriculum. This is making the transition from a purely subject base at post degree or equivalent level to applying this knowledge to a teaching situation in school – it is the start of understanding how subject knowledge is taught to pupils.

If it is appropriate for further written assignments to be set over the course of the training, these should have a firm practical basis to build this understanding. You as a mentor have an important role to play in assisting trainees in their assignments and where possible you should look for opportunities to assist in marking them too.

## Using the standards for QTS in mentoring

There are two kinds of standard relating to knowledge and understanding, which can be broadly divided into 'knowing that' and 'knowing how'. It is useful to think in these terms when mentoring because the 'knowing that' items can be assessed and the 'knowing how' need you to provide opportunities during the school practice for trainees to gain understanding. The standards referred to below are applicable to QTS in England but others are applicable elsewhere in the United Kingdom. The Irish standards for QTS contain a section relating to values; a particularly useful addition.

**Knowing that**

In the secondary phase those to be awarded QTS must, when assessed, demonstrate that they:

- have a secure knowledge and understanding of the concepts and skills in their specialist subject at a standard equivalent to degree level;
- for English, mathematics or science specialists, must have a secure knowledge and understanding of the subject content specified in the ITT National Curriculum (1996);
- for RE specialists, have a detailed knowledge of the model syllabuses.

**Knowing how**

Secondary phase examples include the requirement that those to be awarded QTS must, when assessed, demonstrate that they:

- have a secure knowledge and understanding of the concepts and skills in their specialist subject, to enable them to teach it confidently and securely;
- cope securely with subject-related questions that pupils raise;
- understand how pupils' learning in the subject is affected by their physical, intellectual, emotional and social development.

**Knowing that**

In the primary phase those to be awarded QTS must, when assessed, demonstrate that they understand the purposes, scope and structure of the National Curriculum orders and that they are aware of the breadth of content covered by pupils across the primary core and foundation subjects and RE. They must also demonstrate that they have, where applicable, a detailed knowledge and understanding of the relevant National Curriculum programmes of study and level descriptions.

**Knowing how**

In the primary phase those to be awarded QTS must, when assessed, demonstrate that they understand how pupils' learning is affected by their physical, intellectual, emotional and social development, cope securely with subject-related questions that pupils raise, and know pupils' most common misconceptions and mistakes in the subject.

There is a considerable overlap between primary and secondary phase legislation.

The ITT course should provide opportunities for your trainee to develop specific knowledge and understanding in a number of ways. It is important that you are completely familiar with all aspects of the course so that you can dovetail your mentoring into other aspects of course provision. Look to integrate what you offer in your mentoring into what other members of the course programme are doing. Seek opportunities to develop your trainees' subject knowledge in school wherever possible. It is not always possible for all trainees to join ITE with a sufficient basis for all the subject knowledge they will need to enable them to teach in your school so be ready to support them as they learn and offer opportunities for remedial action.

Questions to ask yourself:

- What qualities and knowledge should would-be teachers possess?
- How would these questions help you decide their potential as teachers?
- How can you tell if a trainee has sufficient subject knowledge to teach?
- How might you find out if an applicant has an aptitude to work with children?

Good practice in developing trainees' knowledge and understanding:

- In selecting trainees, identify balance of knowledge and understanding.
- Don't assume that knowledge will enable trainees to teach their subject.
- Seek ways to identify knowledge and understanding in your trainee.
- Use the TTA's standards for QTS relating to knowledge and understanding.

## Chapter 18

# Planning, teaching and classroom management

In this chapter we shall consider the following aspects of mentoring:

- Helping trainees to plan to teach.
- Planning within the TTA Standards for QTS.
- Helping trainees to teach and to manage a classroom effectively.
- Teaching and classroom management within the TTA standards for QTS.
- Identifying good practice in planning, teaching and classroom management.

### Planning, teaching and classroom management in the TTA Standards

Planning, teaching and classroom management, as any experienced teacher will readily say, are inextricably linked. However, in trainee teachers' eyes – especially in what Furlong and Maynard (1995) call their 'survival stage', when they are starting teaching in schools – there is one dominant focus: classroom management. Good mentoring recognizes this but empowers trainees to look beyond survival, to look beyond tips for teachers, and to plan and teach so that the preoccupation with classroom management alone cannot dominate and skew their view. For a trainee to become a professional practitioner, he or she must understand that classroom management is not the same as control, that teaching is only successful when it engenders learning, and that good teachers do plan – although they may not appear to do so by writing copious notes on pro formas or on paper.

There is a poignant phrase that comes to mind highlighted in the Bath University PGCE course handbook that reads: 'To fail to plan is to plan to fail' – but what do trainees think constitutes 'planning' and is this what their mentors identify with?

We can consider what it might look like to be a trainee planning to teach a first lesson in school. The internal monologue they are playing out in their heads might run something like this:

> The focus is on ME – everyone is watching ME and I have to get my teaching 'right'. I have seen my mentor teach this group and I am very excited to be taking over – it is after all what I will be doing full time in the near future. But I am a bit/a lot scared about so much attention being focused on ME and so I will plan what I am going to do very carefully so it goes well. My mentor has told me what I have to teach and through our mentoring sessions and in my observation of my mentor's teaching I can see how to teach – but will the class like me?

What might it sound like for a trainee planning a lesson later on in the initial training course?

> I don't know what to do. I have spent hours planning good lessons and the class just won't shut up – all they do is chat. I think it's unfair of them – I spent hours planning a really exciting lesson and they didn't appreciate it. I know what I've got to teach – I hope I can get through.

Classroom disruption begins to die down and the trainee is apparently sorting out difficulties:

> Right! I am standing no nonsense from X, Y and Z. Today we are going to do . . . I have to watch T because she tends to go off task – maybe I should see her form tutor again if she does. Wish the scheme of work told me what to teach in this lesson – it's not very clear.

As the trainee approaches the 'plateau' stage identified by Furlong and Maynard:

> OK – I know what I am teaching this week – no problem – I have filled in the lesson plan pro formas ready – last lesson went well – I enjoyed it. We should be through this section of the department's scheme of work in time for assessments at the end of the month. Great!

Several weeks into the initial training course and the trainee is still not 'teaching' and is working in a rather controlling way with individual pupils and with the whole class. There is learning – by the trainee and by the pupils – but it is still not 'planned' learning. This shift cannot occur until the trainee sees teaching and learning as being inextricably linked – the only purpose of teaching is learning and the real focus of planning is to ensure that learning takes place in a continuous and progressive manner. Teaching is interaction and not performance.

If the trainee is to learn to plan, you have to plan so your trainee does this effectively. It is tempting to give 'tips for teaching', to say 'Do it this way and it will work – you've seen me do X, Y, Z and it worked didn't it?' The problem with this approach is that a trainee and a mentor are not and cannot be the same person – and teaching is about engaging between people as well as engaging between subject and teacher and subject and learner – cloning won't work! It can provide some ideas to try out but in the end it is the individual who teaches.

So what might an experienced teacher mean when he or she talks about lesson planning?

Let's imagine it is some time before the experienced teacher is taking on a new class. Because this is an *experienced* teacher there is concentration on content and on creativity – 'how will I teach so that my pupils in this class will learn in a way that excites and interests them?'

Initially the focus is on content: 'What does my class have to know by the end of the year, by the end of each term, by each half term? What is my overview of the content for the whole year?'

Simultaneously the experienced teacher will be finding out about this particular class: 'What do we know about them as a group and as individuals in school? How well do the individuals learn? What did they learn last year? What will I need to note regarding individual pupils as I teach?'

As the experienced teacher engages with planning individual lessons so he or she will turn to seeking creative ways of teaching. This planning takes place in advance of lessons but it happens during the classroom interaction too. Experienced teachers have routines and repertoires and because they know 'what works' in teaching situations they can adjust a plan. Situations can and will arise in class that cannot be planned in advance but, as the teacher is teaching, ideas will come to mind about how to make a particular point clearer or more vivid.

Because experienced teachers have experience of successful planning, their teaching engenders learning and they know that they can 'manage' classes effectively, which gives them scope and space to think about how to teach creatively, to think on their feet and to learn as they do so. The inexperienced teacher usually approaches lesson planning from a very different perspective.

## What does this mean for your mentoring?

You need to be aware that many trainees come to the early conclusion that many teachers do not plan their lessons, or do not need to plan their lessons, because they do not write plans.

- Plan a lesson with your trainee to show what you take into account.
- Explain how to plan in detail and don't assume this is covered elsewhere in ITE.
- Explain why you have planned as you have and invite your trainee to reflect on this.
- Use the pro formas supplied with the course if there are any available – if not, create one.
- Underline the need to look at what has happened/what will happen for continuity.

## Planning within the TTA standards for QTS

As we saw in the previous chapter, the standards tend to fall into two main categories – knowing that and knowing how. The 'knowing that' standards require that the trainee can make a statement that does not have to be drawn necessarily from experiencing the *process* of teaching. The 'knowing how' standards can only be demonstrated in the process of teaching.

## Knowing that (planning)

These standards and sub-divisions of standards can be taught to the trainee – they are perhaps more about assimilating information than achieving mastery in classroom teaching:

- Identifying clear teaching objectives and content, appropriate to the subject matter and the pupils being taught, and specifying how these will be taught and assessed. It is a relatively simple matter to write clear statements of what it is intended to teach in terms of subject content, linking to a class in a year group (but not engaging with the 'character' of a particular class) and specifying, in a few well-chosen words, what will be assessed and the means of assessment – testing, question and answer work, examination.
- Setting tasks for the whole class, individual and group work, including homework, which challenge pupils and ensure high levels of pupil interest. The trainee needs to ensure that in planning statements they make an opportunity for pupils to work together in various configurations – whole class, pair and group as well as on their own.
- Identifying pupils who have special educational needs, including learning difficulties, are very able, or are not yet fluent in English. Knowing where to get help in order to give positive and targeted support. Through writing a planning statement trainees can demonstrate they have had a look through

the class teacher's or special educational needs co-ordinator's (SENCO's) notes on individual pupils, and a reassurance that the SENCO is available for consultation about a particular pupil can ensure positive and targeted support.

- Plan opportunities to contribute to pupils' personal, spiritual, moral, social and cultural development. It is a relatively simple matter to supply a statement of intent at the planning stage ('this lesson will include opportunities for pupils to develop . . . through the following activities . . .'). There is no mention here of the *degree* of contribution to learning in terms of personal, spiritual, moral and cultural development and so the planning might remain at a formulaic statement-of-intent stage.
- Where applicable, ensure coverage of the relevant examination syllabuses and National Curriculum programmes of study. The trainee may be required to reference planning through statements to documentation. So, for example, the lesson plan may give statements of the relevant attainment targets to be covered.

These sub-sections have something in common – the trainee can be told what to include or can see what to include in planning as an observer, without necessarily *experiencing* how teaching itself can and should be in dialogue with, be informed by, and inform planning in each category here.

## Knowing how (planning)

- Setting appropriate and demanding expectations for pupils' learning, motivation and presentation of work. Here the trainee has to demonstrate how their teaching can be judged to be *appropriate* and *motivating*. The expectations can only be experienced as being appropriate and motivating in the context of the lesson being taught and the mentor/observer must be present to record this.
- Setting clear targets for pupils' learning, building on prior attainment, and ensuring that pupils are aware of the substance and purpose of what they are asked to do. Here the trainee has to demonstrate that pupils are aware of what they are asked to do through interaction with them after the substance and purpose of work have been explained. This is asking trainees to demonstrate the effectiveness of their lesson planning.
- Providing clear structures for lessons, and for sequences of lessons, in the short, medium and longer term, which maintain pace, motivation and challenge for pupils. Here the trainee can only demonstrate that he or she knows how to maintain pace, motivation and challenge through the interaction of being observed working with the particular class that the lesson has been planned for.

- Making effective use of assessment information on pupils' attainment and progress in their teaching and in planning future lessons and sequences of lessons. A trainee can be instructed in how to use assessment information in planning but cannot be instructed in how to do this *effectively* – this requires demonstration through the *teaching*.

The temptation when faced with a bank of statements like those in the TTA Circular 4/98 is to approach them in a mechanical tick-list way. This is unhelpful for a trainee, who needs to see that the standards are interrelated and that good teaching is about interaction with learners. In the next illustration we shall meet a trainee teacher who has the potential to be very skilful but she has not yet realized this interaction. Her mentor provides her with opportunities to do this.

---

### Illustration

G is planning lessons carefully and thoroughly, or so it seems. The pro formas that she completes for each lesson show due reference to relevant sections of the National Curriculum, to different groupings for the class: whole, pair, group, and to different types of class activities. The lesson objectives are clear and comprehensive. G states what her pupils should know by the end of the lesson and she also gives a set of objectives for her own professional development. There is appropriate detail in the planning to enable an observer to see why and how continuity and progression are being managed between lessons and, certainly, learning does take place in class. There is assessment and the outcomes of this are used to inform further planning but when the mentor watches this trainee's lessons 'something' is missing.

### Commentary

G is working through the standards for QTS in a methodical and conscientious way. She targets each lesson so that she can show how she is progressing in terms of demonstrating her achievement of individual standards. Her teaching, however, is only *technically* good – it is not dynamically and creatively good. It is adequate, and pupils are doing reasonably well.

The mentor decides that he must move G on – so she engages with the pupils as individuals. What he decides to do is to ask G, before the lesson, to parallel script her plan – when she is doing X what will her pupils be doing? If her pupils do Z what will G do? G realizes that teaching is about more than meeting individual standards – and that she can fulfil these when teaching in a more creative, interactive way that motivates her and her pupils, and retain quality.

---

In the next section of this chapter we move on to consider the second aspect of the standards for QTS as described in Circular 9/92.

## Teaching and classroom management

As we saw at the beginning of this chapter, teaching for some trainees *is* classroom management, particularly during the survival and problem-solving stages. In the early idealism stage the trainee may not even be aware that a teacher does manage a class – the trainee watches the mentor teaching and it may look so effortless and easy. Teacher does X and pupils do Y – exactly as the teacher intended. So it is hardly surprising that trainees suddenly hit a stage where their very identities seem at threat! They try X, just like the mentor, and the pupils do Z . . . not to mention a bit of T (time wasting) and C (stands for chaos).

Most trainees seem to take time to appreciate that experienced teachers have learnt to 'orchestrate' a learning situation – they can plan and teach activities to bring about learning. The performance of teaching is an important stage for any novice to undertake – trying on styles and approaches in teaching for size, to see 'what works and what doesn't' – but the problem is that if teaching remains at the performance stage it is not necessarily bringing about learning – either on the part of the teacher or of the pupils. This is the 'plateau stage' where the teacher and pupils know and do what 'works' at a rather basic, unimaginative level – but there is underachievement on both sides of the learning partnership. The task of the mentor is to move the trainee on and to enable the trainee to move himself/herself on.

> There is a temptation to divorce planning, teaching and classroom management and simply describe best practice in each. Good teachers produce well planned lessons where teaching is effective and classroom management is not an issue as the activities pupils are engaged in are interesting.
> (Nicholls, 1999: 5)

This scenario would be a rather simplistic view of good teaching. Trainees can produce good plans with interesting activities that potentially interest pupils and still they do not teach well!

> It would be easy to look at the standards for QTS and the ITT National Curricula subject knowledge requirements and assume that secure subject knowledge and some practical class experience will necessarily lead to quality teaching.
> (Nicholls, 1999: 8)

Good teaching empowers the learner not only to learn what is intended but to be inspired to enhance that knowledge through further study. Teaching is more than knowledge – it is empowerment to learn how to learn. If the sole objective of teaching is learning then the sole objective of classroom management is the promotion of learning. Teaching is not synonymous with education – education is fundamentally beneficial – but teaching may be beneficial or not.

What do novice teachers understand as teaching? They tend to think that learning happens as a result of teaching and – often rather simplistically – they think that learning has taken place because something has been taught. If only education were so easily managed! In the early idealism stage they see teaching and learning occurring in the classroom, usually the classroom of an experienced and able teacher, and the process looks straightforward. Teacher teaches X and pupils automatically learn X – thus they have been educated in X! Were it so simple!

The experienced teacher knows that learning does not just happen – it is managed by the teacher, and even then it can be hit and miss. In a sense it is because novice teachers see such able practitioners that they think teaching IS just giving information. Where they see problems in a lesson they tend to assume at first that it is the pupils who are being awkward – and this reinforces their initial belief that the work of a teacher is to survive in the hostile world that is the average school. It is only later that trainee teachers come to realize that even the best teachers have lessons where they struggle and that sometimes this is because they are not teaching well!

The novice who has been observing a very capable and experienced teacher is usually shocked to find that just giving the information to pupils in the same way as the usual teacher has done isn't working. At this stage panic can set in – the trainee is doing everything 'right' in copying the class teacher and learning isn't happening! The most obvious sign that learning isn't happening is that the class is being disruptive – and challenging the teacher – either by confrontation or by apathy. As good teaching *seems* to be good presentation, and because the trainee is only really aware of being a teacher and not of what is going on inside the learners (or potential learners) in front of him or her, lesson planning focuses on input from the teacher.

Lessons are almost entirely teacher led (when under attack you don't lend ammunition to the enemy in front of you!). Where trainees often see teaching as being particularly problematic is establishing an appropriate professional 'distance' in their work with pupils. The difficulty of establishing an appropriate distance is exacerbated by watching a teacher who has already done the spadework with a group – and does not have to struggle for control. In this context the 'distance' between teacher and class may appear close but it is a closeness born of knowing and respecting personal and professional boundaries between learners and teacher.

Trainees begin to emerge into the 'identifying difficulties' stage when they realize through experience that teaching is about problem solving. Teaching is a transaction where the contract is that learners take in the information given by the teacher and process it internally. Good teaching occurs when learners feed back their learning – regenerating the teacher – through knowledge (moving the teacher's thinking on) and through positive personal reaction too. One reason why trainee teachers find teaching so difficult and exhausting at first is that they are simply drawing from their own reserves of knowledge and physical and emotional resilience.

Juggling different groupings – whole class, groups and individuals – is difficult enough because it feels like the manageable mass will at any moment run amuck! It takes courage to vary the arrangement of learners in a class by breaking the mould of 'teacher leads class and class follows'. Often, personal teaching objectives overshadow learning objectives to such an extent that they can be eclipsed. My objective for this year 10 lesson is to survive!

Trainees move on when they see that there may be solutions to problems but that they are not ready made. It is difficult for anyone to have to learn that there may not *be* an answer – or certainly a universally useful one. For trainees putting their reputation on the line each time they teach, it can be doubly frustrating and confusing. It is tempting for the mentor to 'tell the trainee how to teach' and, to a certain extent, this 'training' is a part of enabling a trainee to become a teacher – not everything in teaching can or should be left for the trainee to 'discover'. They need opportunities to develop their own style of teaching, and sufficient time to do this; trainees are learners too and they need room to make mistakes and to experiment if they are to progress.

Do not forget what it is like to be a raw beginner – especially if you have been teaching for some time. If we carried round all the bruises and pain of learning as well as all the excitement, wonder and joy of it, we would never survive as teachers – we need to be realistic optimists. As it is difficult to explain every nuance in teaching, it is tempting to tell trainees 'do A and B will follow'. When trainees come to teach and to learn how to teach, it is important to explain to them that there is no box of goodies called 'good teaching' although there are strategies and techniques that will help. If trainees do not have the potential to teach, and this is a personal as well as a professional aptitude, no amount of teacher training will enable them to become good teachers.

We shall now look at how aspects of teaching and classroom management are set out in the TTA's standards and, once again, we shall consider them in terms of 'knowing that' (acquiring information) and 'knowing how' (through experience).

## Teaching and classroom management within the TTA Standards for QTS

Teachers need to be familiar with the code of practice on the identification and assessment of special educational needs. Only this statement of the standards relating to teaching and classroom management is about gaining information rather than experiencing practice. The remaining standards in this section can only pertain to actually being involved in the process of teaching the 'knowing how'.

But how can trainees be encouraged to develop their teaching if they cannot be taught to teach? The answer lies in asking appropriate targeted questions. Trainees need to be supported as they develop their teaching skill, but if they are to be given the context and the support in which to develop, the mentor and the trainee need to ask and answer some key questions. Looked at from this perspective it is possible that some trainees cannot be developed in their teaching in some schools. The *context* of teaching, which includes the teaching environment – the age and range of classes observed and taught as well as the staffing involved – has to be appropriate for them to develop the skills they need to be 'good' practitioners. Some trainees will learn better in one kind of environment, others will flourish in another. What *is* important when using the standards for QTS relating to teaching and classroom management is to ensure that the trainee is placed in a department and a teaching situation which is appropriate to their development needs.

## What does this mean for your mentoring?

- Ask as well as answer questions relating to teaching and classroom management.
- Stress that good classroom management results from good planning and teaching.
- Share strategies for classroom management and ensure the trainee uses them.
- Provide opportunities for your trainee to experiment and not just teach 'safe' lessons.

In the final illustration in this chapter we see that a trainee can fulfil the statements relating to subject knowledge and understanding as well as those relating to planning, teaching and classroom management. Even so, her teaching has not been as inspiring as it might have been.

## Illustration

M is a very well-qualified English teacher. She is working in a school where the emphasis in the subject department is on teaching grammatical correctness rather than fostering creative writing. She draws from her own considerable knowledge base for planning her lessons but is encountering difficulties because she is creative in her approach to teaching her subject and finds the grammar-based approach limiting and frustrating. Her preferred style of teaching is to encourage the pupils to interact in a freely expressive way but she needs to develop sufficient classroom management techniques to enable her to do this without pupil disruption. How can she gain this experience in her present school where this is not the preferred way of teaching? Her mentor quickly identifies the problem and between them they come up with a useful solution. M will learn how to teach in this environment but she will have the opportunity to visit a neighbouring school – where the mentor has a friend and where the atmosphere is more relaxed and less formal. Given this opportunity, M learns how to teach in her preferred style.

## Commentary

Here the needs of the trainee are put above the immediate needs of the mentor and the main school placement. The trainee needs to develop a particular kind of teaching skill if she is to feel secure and fulfilled. By observing teaching in this style and by accessing opportunities for team teaching she has the best of both worlds. Now she can see the benefits of both traditional grammar-based and creative teaching and comes to the happy conclusion that good teaching embodies both approaches and is best tailored to meet the needs of individual pupils.

Questions to ask yourself:

- How can I work with my trainee so we plan, teach and manage together?
- How can we work together to improve our planning, teaching and class management?

Good practice in planning, teaching and classroom management:

- Show by example, sharing and explanation how good teaching is the result of thorough planning and effective classroom management.
- Ensure your trainee has opportunities to observe a range of successful approaches.
- Develop your trainee's awareness that good teaching is reliant on interaction in class.

Chapter 19

# Monitoring, assessment, recording, reporting and accountability (MARRA)

In this chapter we shall be considering the following aspects of mentoring:

- MARRA is at the heart of good practice in teaching.
- Aspects of MARRA are interrelated but have different training implications.
- MARRA is concerned with personal and professional responsibility.
- Identifying good practice in mentoring for MARRA.

## Monitoring, assessment, recording, reporting and accountability

This is arguably the most difficult area of the standards for QTS that your novices encounters. The reason they find it difficult is that it is so complex, as the multiplicity of themes within the title suggests. The rhetoric is impressive; the practice sometimes less so. The other main reason lies in the need for trainees to be teaching at a level beyond the 'survival' stage in order to appreciate the significance of each of these five interrelated aspects of teaching. Not surprisingly, MARRA, as it is affectionately known, can present considerable problems for even the most experienced teachers. This is why it has become a prime focus of attention within our OFSTED inspections.

Each part of MARRA is interrelated because:

- Without learning there is nothing to monitor.
- Without monitoring there is nothing to assess.
- Without assessment there is nothing to record.
- Without recording there is nothing to report.
- Without reporting there is no means of being held accountable.
- Without accountability there is no motivation to learn.

It takes time to manage MARRA effectively but it is not something additional to classroom teaching, it has to be pivotal to good teaching. What is the point of teaching a lesson if at the end of it we cannot be sure how much has been learnt by pupils or how much needs additional teaching? What is the point of teaching without assessing, if not to see what to teach next? What is point of assessing if we fail to record the outcomes of assessment to inform others and ourselves? How else can we report on individual pupil's progress and review the effectiveness of the scheme of work? We *are* all accountable, even at the trainee stage in teaching, so let us enable trainees to approach MARRA in a responsible way that informs rather than constricts. The problem that we mentors face is one of encouraging our trainees to comply creatively in meeting competence statements that cannot represent all there is to being a professional educator.

For trainees, the temptation is to see monitoring as just 'keeping an eye and making sure pupils are busy', to see assessment as just 'tests' and examinations, and recording as a neat row of marks out of 10 that do not inform but merely record. The point of MARRA is to consolidate and extend teaching and by doing so to support learning!

## Monitoring

How do I explain the significance of monitoring to my trainee? Maybe with a metaphor:

Imagine a heart monitor attached to a patient in hospital. It flickers constantly as it adjusts the readings of the heart, but it takes skill to interpret. Monitoring without the knowledge to interpret what the indications signify is a waste of time. So now let us imagine that a teacher sets a piece of written work for pupils – they hand it in for marking. The teacher looks at it and writes a variety of grades on it – perhaps between 1 and 10 – and a comment – good, satisfactory, poor. The pupil looks at the mark and at the comment and thinks 'I got an X out of 10; that makes me an X out of 10 student, I got good/ satisfactory/poor' – and then just puts the book away . . .

A ridiculous scenario? Perhaps, but one that has occurred in countless classrooms for countless years and will be replicated until all novice teachers are taught how to monitor progress effectively. They need training in how to develop expertise in giving feedback to inform learners through marking. Unless they are shown how to 'mark and inform pupils' assigned classwork and homework' and how 'to enable pupils to progress through setting targets and action plans', pupils' learning will be curtailed.

Trainees need tuition from you to support them in:

- attuning constructive and informative feedback to monitoring pupils' work;
- monitoring that this marking results in demonstrable progress by pupils;
- learning how and when to check that work set has been understood and completed.

Trainees may remember regular tests from their own schooling, but will they be aware of how a teacher monitored progress? Did their teacher necessarily monitor their progress as a basis for informative feedback to improve their learning? We cannot assume that there have been effective role models for our prospective teachers that they can emulate and so their training for monitoring needs to be overtly comprehensive. The tests trainees underwent as pupils were used only to show the teacher what strengths and weaknesses a pupil had – rather than as 'a basis for purposeful intervention in pupils' learning' to move them forward in their learning.

But how do we train teachers in a wider repertoire of activities than setting tests? Trainees need to learn the art of the targeted question so you may need to give tuition and practice in asking the right question at the appropriate time for a particular pupil to show their learning. Targeted questioning is an art that takes time, patience and skill for a trainee to develop, not to mention time, patience and skill for you as mentor to work with your trainee!

An effective way of seeing how far your trainee is learning to monitor pupils' learning by targeting questions is for is for you to model targeted questions in mentoring:

- When you teach, how do you monitor your pupils' work?
- In this activity, which questions will you ask to monitor progress?
- How will you differentiate your questions to monitor individuals' progress in class?
- How will you ensure that you monitor every individual's progress in a lesson?

- Apart from setting tests in each lesson, how do you know what each pupil has or hasn't learnt and how might you record this ongoing monitoring in your mark book?

One of the activities that many mentors find useful in enabling their trainees to learn how to monitor attainment of individual pupils is to model good practice for the trainee. In this first illustration we shall we how one mentor sets about doing this.

---

## Illustration

J is an experienced mentor. She has two trainee teachers working with her. One understands how to monitor pupils' progress and to use her observations to move them on – almost from the outset of training. The other doesn't. J uses several approaches to enable this trainee to learn:

- She models how to monitor pupils' progress through team planning prior to team teaching. Before the lesson she explains to the trainee how they both will monitor the progress of individual pupils by checking pupils' books as they walk round during a writing activity and what questions they might ask.
- She asks the first trainee to join them for a mentoring session – the trainees work together as peer mentors – 'I do X' to check that pupils have understood.' 'I do Y. . .' They learn from one another as they learn to teach.
- She sets a monitoring task for both trainees. After they have taught a lesson each she asks them to look at the mark book for each class. She picks out the names of several pupils in each class list and asks how that child was working in class. How did each trainee find out how pupils were progressing?

## Commentary

There is much to commend J's practice as a mentor. She enlists the expertise of the peer trainee to show that it is possible even during training to assess and monitor progress as well as to deliver a lesson. She models how monitoring is part and parcel of the process of teaching. By targeted questioning about particular pupils in the class list she shows both trainees how the mark book must be a tool for checking learning.

Finally she trains the mentees to use the mark book effectively – to record pupils' progress in a systematic and informative way, to enable her to plan subsequent lessons and she links the marks to the National Curriculum levels on a regular basis to make reporting easier.

---

## Assessment

According to the current standards for QTS, trainees must demonstrate that they can assess how well learning objectives have been achieved and use this assessment to improve specific aspects of teaching. The key word is teaching. Thus assessment informs *teaching* and learning assessment means more than a regular setting and systematic marking of tests and examinations. Trainees need to be taught the difference between setting tests and examinations in a summative capacity – at the end of a programme of study for example – and setting them to *inform* ongoing learning. Trainees need to appreciate that planning must include opportunities for pupils to learn how to learn as well as moving them on to a new programme of work. Because they are expert learners, trainees sometimes overlook this 'learning to learn'.

They also need to learn to appreciate the benefits of self-monitoring and assessment by pupils, which leads to autonomy, ownership and taking on additional responsibility.

According to current government legislation, which is set out in the requirements of Circular 4/98, trainees must know and must demonstrate that they:

- know statutory assessment requirements;
- where applicable, understand the expected demands of pupils in relation to each relevant level description or end of key stage description and, where appropriate, the demands of syllabuses and course requirements for GCSE, KS4 and post-16 study;
- where applicable, understand and know *how to* implement and assess requirements of current 14–19 qualifications;
- assess pupils consistently against attainment targets, if applicable.

The direction for mentoring practice in 4/98 is clear: 'trainees may need guidance from an experienced teacher in assessment of work against attainment targets.'

The implication is that, where it is appropriate for trainees to be assessing work against attainment targets, knowing that there are different levels is not enough. Trainees need to be mentored so that they know how to assess against the National Curriculum and to know how, they must work alongside mentors who do this too!

In the next illustration we shall see how one mentor has set out to ensure that this occurs.

---

## Illustration

M is a mentor in a secondary school department where the attainment targets are not used as a basis or adjunct for recording pupils' marks in mark books. Her colleagues say there is barely time to teach let alone work out what each piece of assessed work means in terms of attainment targets. Although marks are recorded consistently by all members of the subject departments and there is some opportunity for moderation between teachers' marking during the year, M realizes that something has to change. She will need to demonstrate to her trainee that she not only understands the National Curriculum requirements but that she uses the attainment target (AT) levels to inform her own planning and teaching. However, she is bound by departmental policy and does not want to upset her colleagues or make an unmanageable amount of work for herself or others.

She decides to use the AT levels in a formative way herself and, if it is successful, to raise awareness in departmental meetings at a later date. First, she looks at the statutory requirements for the National Curriculum. The end of KS 3 requires that pupils' progress be assessed against AT levels over a range of work and over a period of time. In planning tasks for her class she jots down in her mark book against the date the highest AT level that she feels her pupils can attain in this particular exercise. Realistically she cannot do this for every exercise in every lesson, so she chooses key activities within the scheme of work to assess on a regular basis. When she collects in marks for each individual pupil on a weekly basis she notes the AT level approximates to that each pupil is working at. There are two lines of marks in her mark book. One records raw scores in tests – one on a weekly level records attainment target levels. Having got herself accustomed to using AT levels in planning and monitoring achievement she introduces this system with her trainees.

## Commentary

M has no difficulty training her mentees in the requirements of meeting examination syllabuses and in showing them what assessment means in end-of-year examinations. What *is* difficult is meeting the trainees' needs in learning to use National Curriculum Attainment Target levels in their marking where this is not departmental policy. She pilots a system for using AT levels and uses her own experience to enable her trainees to use AT levels too. She hopes to introduce the use of AT levels at departmental level, once she has evidence that this raises standards in pupils' work in her classes. In assessing pupils' work she tells them what the different levels mean in pupil-friendly terms. They are learning to recognize what they must do to progress to higher levels. She is effectively educating herself and her pupils as she educates her trainees.

---

## Recording

How can a busy teacher adequately record what happens in one lesson so that it forms the basis of planning subsequent lessons and enables clear reporting on individual pupils' progress to parents and to colleagues? No one could pretend this is easy – but it *is* possible – and it is essential to good practice! When trainees are given a mark book they need to be trained in how to use it – effectively! They need to know *why* they should use it – and not just because their mentor says so. They need to understand how the recording of marks can make their job easier and more effective. There is no one accepted way of using a mark book well – practices differ so widely between schools that trainees miss out on an amalgam of good practice as they undergo training across different schools and can be left confused and unsure about recording progress. Once they can understand that teaching is to enable learning – theirs and the pupils' – they can appreciate the mysteries of the ubiquitous mark book!

Let us focus on one statement in the MARRA requirements of Circular 4/98:

> those to be awarded Qualified Teacher Status must, when assessed, demonstrate that they . . . check that pupils continue to make demonstrable progress in their acquisition of the knowledge, skills and understanding of the subject.

The key word again is 'demonstrable', as recording makes demonstrable what is being learnt in knowledge, skills and understanding and if it doesn't it is a waste of time! Notice that it is not just *knowledge* that must be demonstrably assessed but *skills* too. The first stage in preparing to mentor for MARRA entails making sure that, as an expert, you know what are the skills as well as the knowledge that pupils must master in your subject area – so that you can enable your trainee to learn this too.

A mark book used properly will inform, not just record. It is useful to have a section not just for recording marks, but to jot down comments on a regular basis about each pupil's progress in terms of subject skills. This could be coded into a shorthand of symbols to indicate progress so that every month, for example, there is an overview of developing skills. Assessment can and should be used for different purposes – for National Curriculum and other standardized tests and baseline assessment where relevant – and its outcomes need to be recorded in a way that can inform planning by the teacher and by pupils as they are educated into becoming autonomous learners.

---

## Illustration

L identifies the skills and knowledge that her class will need to master in the next section of the subject programme in the scheme of work. She works with her trainee by sharing perspectives as they analyse the next part of the teaching programme. They decide what it is that pupils must be able to do and know by the end of the next section of the syllabus. Having done this they decide how this should be recorded in mark books – how will they know and be able to demonstrate that pupils know the necessary skills and knowledge that the syllabus requires that they know?

## Commentary

This mentor recognizes that recording is part of planning – planning how to record so as to inform further planning and further teaching to bring about better learning. Planning needs to include making an opportunity to record as well as to consider data resulting from it. By planning the skills and knowledge that will be assessed and how the outcome of assessment will be recorded, L is planning ahead to report. She is also giving her trainee the opportunity to be well prepared for the accountability that will ensue later in the year in writing reports to parents. Gone are the days of the 'must do better' statement. In this approach, the teachers clarify the skills and knowledge so they can be made clear to the pupils. The pupils are told not only what they are expected to learn within a lesson but why; not only how they are doing in a particular subject but how to improve. The key is to identify what is to be assessed and how to record it well.

---

## Reporting

In the following illustration we have an account of how one teacher learnt the importance of effective recording – a hard lesson but one that spurred her to improve her practice.

---

### Illustration

There were two boys in my year nine classes who shared the same first and family name. One was a dual linguist, the other a pupil with special educational needs who struggled to master even his own language. I assessed their progress regularly so I had a sound understanding of their ability in terms of skills and knowledge but somehow I mixed up their reports during my preparation for parents evening. Imagine how my heart sank when, having said how delighted I was about X's progress in French and Spanish, a very kindly dad explained that his son took just one language, French. I had failed to use my recording as a basis to inform my planning for reporting back to pupils' parents.

### Commentary

Had this teacher looked at her mark book and at the class lists when preparing for parents' evening she would have spotted her mistake. She thought she knew which parents went with which pupil and how each was progressing. Instead of making doubly sure by checking that she was talking to the appropriate set of parents, she assumed she was doing so. Sometimes we learn more about good practice from mistakes than from simply emulating glowing examples of good practice. The moral of this illustration lies in avoiding making assumptions when reporting back to others!

---

In the second illustration we return to looking at an example of good practice as a possible model of how to include trainees in the process of learning how to report.

## Illustration

M is the professional tutor in school and organizes the trainees' programme. In consultation with staff undertaking mentoring, it has been agreed that all trainees will shadow their mentors in writing reports and will attend parents' evenings. Parents are asked if they mind if a trainee is present and, rather like the situation in training hospitals, they are informed that a trainee may be present during the interview. They can refuse and ask for a meeting with the teacher without the trainee. Because the trainee has not only observed the report-writing process but has been involved in writing the report in conjunction with the mentor, he or she is in a strong position to contribute appropriately during parents' evenings. This raises the status of the trainee in the eyes of pupils and parents. Learning to report on progress is 'for real' within training.

## Accountability

Whether we like it or not, we are in an era of increasing accountability in education. The reliance on standards and competences as well as the mushrooming of public accountability within the media – of politicians, doctors and teachers – evidences this. 'Name and shame' predominates and needs to be counterbalanced by a healthy 'name and value' cultural shift. As teachers we are held to public accountability where we need to show what we *do* achieve, sometimes against the odds. Where there are weaknesses we need to address them in a climate of support, not isolation.

New entrants to the teaching profession know that they are constantly accountable. The spotlight never leaves them as they seek to amass evidence to support their claims of having achieved against statements of competence that indicate their readiness to teach. They hold the heavy responsibility of proving their readiness in a way that has not been seen before. Standards for QTS, standards for ICT, Standards for induction, Standards for subject leadership . . . – so much accountability, so much need for support and challenge to succeed. Pupils are accountable for their work to teachers, teachers are accountable for their work to government as it uses national, local, comparative and school data to set clear targets for pupils' attainment and for improvement in teaching standards. But accountability is often seen as an externally imposed mechanism – a bureaucratic exercise that impedes.

How can we, as mentors, enable trainees to use accountability as a vehicle for enhancing our practice and that of other teachers, and for our pupils' progress in school? It is crucial in a climate of accountability to take personal responsibility not only for achieving against others' standards but against those we set ourselves. We have a duty to ourselves to retain and develop a vision of what good teaching is and to share that vision with new entrants to our profession. The key is for more experienced members of our profession to drive accountability rather than to be merely driven by it. There is a danger that statements of competence will determine initial training programmes. Perhaps our greatest challenge as mentors is to provide opportunities for trainees to develop their creativity and retain vitality while meeting accountability

Using OFSTED as a basis for improving practice is a case in point. The provision of ITT in schools is now inspected as a routine part of the process of accountability of ITT providers to the government in the UK. Trainees will often encounter OFSTED inspections during their period of training or within a short time of achieving QTS. As mentors we owe it to trainees to prepare them to be accountable, to give them opportunities to build their confidence. We need to share a vision that accountability has benefits and to see opportunities for their development in accountability, rather than conditioning them to expect to be ground down by it.

---

## Illustration

The OFSTED inspection of a subject area in a PGCE course is announced. During the inspection the HMI will visit several schools to look at how well trainees are being prepared for entry to the profession of teaching. The HE-based subject co-ordinator calls a meeting for all mentors. He begins the process of preparing colleagues in schools for what will be a protracted process of inspection – not a one-week visit as it is in school inspection but a series of visits and feedback sessions between November and June. When the forthcoming inspection is announced there is no way of knowing which school the inspector will choose to visit or when, and so everyone is asked to attend. Knowing that the focus of inspection will be on how key areas for development have been addressed in schools and in HE the discussion centres on MARRA, the weakest area in the previous report on the PGCE programme

J is new to mentoring. She works part time and she has not been in her school very long – and she *is* aware that her own practice in MARRA needs upgrading. She asks her HE colleagues to organize a further meeting and to invite mentors who are successful in MARRA to contribute. During the meeting she listens to J who has developed her own highly successful way of ensuring that trainees remember to use MARRA well. She trains them to think in terms of learning to drive – mirror, signal, manoeuvre!

MIRROR. Before teaching something new, look back at what has gone before and check that it is safe to proceed.

SIGNAL. Let the pupils know why they are doing something new and let other colleagues know too – through fully completing the lesson planning pro forma for example.

MANOEUVRE. Having looked at what has been done before and confirmed that it is safe to move on, the trainee must look ahead to predict what will happen – just as a driver does (or should do) before engaging in a course of action.

Mirror – signal – manoeuvre! The art of safe driving – and good teaching.

## Commentary

This good practice is all about creative compliance within accountability! Everyone involved in this ITT partnership is accountable to the inspection. All mentors are invited to attend the pre-OFSTED meeting. During the meeting one of the mentors realizes that she needs to upgrade her practice. She realizes that she is likely to be selected for inspection because she is new to mentoring and she seeks help. She targets help from her other mentors and she and the HE tutor can learn about good practice in MARRA.

Monitoring, assessment, recording, reporting and accountability are becoming the driving forces in our profession. We need to enable trainees to realize that, while this is inevitable, it does not necessarily mean an end to creative, imaginative teaching.

Questions to ask yourself about MARRA:

- How do I use the different forms of assessment (formative, summative, criterion referenced, norm referenced, diagnostic, formal and informal)?
- How far does MARRA enable me to ensure continuity and progression in my planning for my teaching?
- Do I know what the Autumn Package is and how to use the information in it?
- Am I using national test data in a diagnostic way to improve pupils' work?
- How do I monitor individuals' progress against the attainment targets in the National Curriculum? What use do I make of my monitoring in reporting?
- How does my mark book enable me to keep tabs on individuals, as well as the whole class's progress, across a range of work and across a period of time?
- What opportunities do I provide to encourage self-assessment by pupils?
- What are the areas of MARRA that my trainee and I need to work on together so we can both improve what we do?

Good practice in mentoring for MARRA:

- Trainees need practice in learning to target questioning for pupils.
- Trainees need to ensure that their assessment drives planning their lessons.
- Trainees need tuition in linking their mark book to the National Curriculum.
- Training needs to include opportunities for supervised reporting to parents.
- Trainees need to realize their accountability but retain creativity and vitality.

Chapter 20

# Other professional requirements

In this chapter we shall be looking at the following aspects of mentoring:

- Providing your trainees with access to legislation relevant to teaching.
- Enabling your trainees to understand how duties frame their teaching.
- Accessing sources for continuing professional development.
- Identifying good practice in being a professional educator.

We have a responsibility to ourselves as teachers and to our pupils to ensure that our schools are learning organizations, by developing a culture where not knowing is accompanied by the commitment to improve by finding out and sharing knowledge.

What else could there be to teaching beyond knowledge and understanding, planning, teaching and classroom management, monitoring, assessment, recording, reporting and accountability? What do the current Standards for QTS say? What do they not say that we, as professional educators, know they should? How do you measure passion for teaching? How do you measure vision and life-affirming energy? We know these qualities when we meet them in our work with trainees – and we certainly know when they are missing! What *is* it that really defines a good teacher? Let's start with a look at current legislation for QTS and then reflect on what cannot be 'measured' but is no less a 'professional' requirement.

Trainee teachers need to know the implications of being a teacher. Teaching, as we mentors know (although some trainees do not know it when they begin training) is much more than being with pupils, more than engaging with learners. Teaching is bound by traditions, conventions and rules. As mentors, we have a responsibility to guide our trainees through the maelstrom of legality and obligation and responsibility and ensure that they retain their enthusiasm and commitment for effective, creative teaching.

If we want to develop true artistry in teachers' work in the classroom and raise public awareness of this we must first explain the ground rules to our would-be teachers.

## Providing your trainee with access to legislation relevant to teaching

Copies of relevant documentation are sent out to schools and should be made available for all the staff to consult. The problem here is that documentation tends to sit on shelves until it is needed – in emergency. For a school to undertake involvement in ITE there is an added impetus to ensure that all staff, teaching and non-teaching, know where the documentation is held and the implications of that documentation for working with pupils in school. Ignorance is not bliss or an excuse.

How can you, in your mentoring, ensure that you do know and understand current legislation? One effective way is to read the relevant documentation and reflect on its implications for your teaching and then to share your reflections with your trainee. But how do you manage to access this documentation? How do you ensure that you are up to date? The answer is becoming increasingly simple: you can access legislation that relates to teaching, and which trainees need to know about, via the Internet.

You are likely to find the following sites particularly useful:

http://www.open.gov.uk

This site relates to the present parliament and sets out recent legislation.

http://www.ngfl.gov.uk

Perhaps one of the most useful sites for teachers, this is the address of the National Grid for Learning and from it you can access government agencies and departments and gain access to resources of the leading educational agencies in the UK, including Becta, QCA and TTA.

## Government agencies

- British Educational Communications and Technology Agency (Becta).
- Qualifications and Curriculum Authority (QCA).
- Teacher Training Agency (TTA).
- Northern Ireland Council for Curriculum Examinations and Assessment (CCEA).
- Qualifications, Curriculum and Assessment Authority for Wales (ACCAC).
- Scottish Consultative Council on the Curriculum (SCCC).
- Scottish Council for Educational Technology (SCET).
- Scottish Qualifications Authority (SQA).
- Centre for Information on Language and Research (CILT).
- Public Record Office (PRO).

## Government departments

- Department for Education and Employment (DfEE).
- Special Educational Needs (SEN) division.
- Choice and Careers Division.
- Careers and Occupational Information Centre.
- Department for Education for Northern Ireland (DENI).
- Scottish Executive Education Department (SEED).
- Scottish Executive Enterprise and Lifelong Learning Department (SEELLD).
- National Assembly for Wales.
- Office for Standards in Education (OFSTED).

From the above list of agencies and departments it is clear that, as a mentor, you can have fingertip access to legislation contained within the standards for QTS. The major education sites have an added facility that makes them invaluable tools for accessing information, not only for you but for trainees. You could set a trainee the task of accessing information about a particular statute.

Within the standards for QTS teachers have legal liabilities and responsibilities relating to:

- the Race Relations Act 1976;
- the Sex Discrimination Act 1975;
- Section 7 and Section 8 of the Health and Safety at Work etc. Act 1974;
- what is reasonable for the purposes of safeguarding or promoting children's welfare (Section 3(5) of the Children Act 1989);

- the role of the education service in protecting children from abuse (currently set out in DfEE Circular 10/95 and the Home Office, Department of Health, DfEE and Welsh Office Guidance *Working Together: A guide to arrangements of inter-agency co-operation for the protection of children from abuse*);
- appropriate physical contact with pupils (currently set out in DfEE Circular 10/95);
- appropriate physical restraint of pupils (Section 4 of the Education Act 1997 and DfEE Circular 9/94);
- detention of pupils on disciplinary grounds (Section 5 of the Education Act 1997).

There are several ways in which you can access details of legislation that are relevant to teaching. It may be that the HEI provides the trainees with details of relevant government documentation or it might be appropriate to photocopy relevant sections to give to the trainees to study. But study is not enough by itself. The trainee needs to understand how legislation impacts on practice and how theory informs teaching, and this is where you have a role to play as a mentor by explaining these implications. Non-contact days in school are an under-used resource in schools (according to Harland *et al*, 1999).

A useful focus might be to ensure all staff are aware of their duties under the law:

> those to be awarded qualified teacher status should, when assessed, demonstrate that they are committed to ensuring that every pupil is given the opportunity to achieve their potential and meet the high expectations set for them.
> (Circular 4/98)

Trainees need to realize that, although teaching does not guarantee pupils' learning, it can and should provide opportunities so that all pupils have access to effective learning. For learning to occur, teachers need to synthesize skills and knowledge and be able to ensure the equality of opportunity among all pupils within their care. For trainees to appreciate this synthesis, mentors need to be truly 'professional' in their conduct. As mentors we are role models not only for our pupils but also for new pupil-educators.

As the next illustration shows, some trainees can seriously misjudge their image and their responsibilities as a role model and as mentors we have a responsibility to take firm immediate action. It is our profession that it is a stake, not just one school. We have a joint responsibility to one another to define what we know to be professional.

## Illustration

Nobody will ever forget that day! G arrived in school to teach her year 10 pupils. 'Make sure you interest them by showing you value what they do – like clothes for example', G had been briefed by her mentor. At the moment she arrived – or rather made an entrance – the mentor was deep in phone conversation with the university tutor. The conversation suddenly hit a hiatus. There she was, our trainee, regaled in skin-tight, lime-green hot pants and black fishnet tights – the pupils were interested . . .

'Oh my . . .!' 'You'll *never* believe what . . .' 'I have to go *now* . . .'

Effectively ambushed by the mentor and the professional tutor, she was dispatched home to change into more professional clothing – without a moment's hesitation.

## Commentary

G's mentor was quite right in telling her to find an approach that would interest her pupils and it seems incredible that a trainee teacher could make such an elementary mistake in the dress code. But trainees are sometimes nascent professionals and we, as mentors, must guide, support and work alongside them, sometimes in ways that seem so very obvious to us. It is *our* profession and the standards we are judged by come from our everyday work. We cannot apportion blame to 'someone' else if professional standards are infringed, because we have a shared responsibility to evolve standards.

This illustration offers us another important perspective. Professional conduct is about good teamwork. This is the essence of good teaching and of effective schools. Governors, parents and agencies beyond the school work in unison to bring about an environment where pupils can reach their potential. Teaching unions have a leading role in representing and maintaining the standards of what it is to be 'a professional'. It stands to reason that trainees must have a working knowledge of this intricate network – mentoring must involve opportunities for trainees to look at the classroom from other perspectives than that of a teacher. Through some devolved mentoring to a governor or perhaps to a union representative, a mentor can ensure trainees learn about the work of professional colleagues who support their teaching within a school.

A further effective way of doing this is to involve trainee teachers in out-of-class activities. They can learn much about the interaction between school and other professional agencies by participating in a residential trip away from the school site.

## Illustration

D is learning to be a teacher in a school where the new entrants – year seven pupils – go away for a few days together near the beginning of the school year. D is invited to go for part of the stay with the children so she can appreciate how bonding occurs.

## Commentary

D is closely involved in the process that leads up to the school visit because she is working for some of her time with the head of year and has a tutor group that she shares with her form tutor–mentor. This brings her into contact with parents who write in to ask questions about the visit and with care agencies, for some of the group are in the care of the local authority. It brings her insight into how her tutor group is a part of a network that supports the school community and involves non-teachers.

---

There is a parallel to be drawn between the statement within Circular 4/98 that we focused on earlier in this chapter and the needs of the trainee teacher in school:

> those to be awarded QTS must demonstrate that they are committed to ensuring that every pupil is given the opportunity to achieve their potential and meet the high expectations set for them.

A similar obligation is implicit in the mentor's work. For trainee teachers it is the responsibility of their mentors to ensure that 'those to be awarded QTS have the opportunity to demonstrate that they are committed to ensuring that every pupil can achieve their potential.'

The problem about setting down statements of competence is that they tend towards measuring that which is easily measurable and there is a likelihood that in doing so we ignore what is not easily measured as a requirement, even when it is essential. Creating opportunities for mentees to learn about 'other professional requirements' of teaching is a responsibility that all involved in ITE share. Teaching is a privilege. There is surely no greater gift for a person than to mould the conduct and inspire the minds of children through their work and their nurturing. The word 'education' has its origins in animal husbandry and to engage in teaching is to engage in rearing. Not surprisingly, it can give rise to close personal as well as professional relationships.

The line between personal and professional can be difficult to determine – especially for a novice. Trainee teachers do not have full legal responsibility for pupils until they are qualified and yet they can find themselves acting in *loco parentis* – there may not be someone else to turn to when a crisis occurs in school. It stands to reason, therefore, that mentors must prepare mentees for just such a crisis. Part of mentoring is taking time to ensure the trainee knows what to do and who to turn to in an emergency. Just how close is appropriate within a professional relationship? How close is too close when working with individual pupils or in class needs to be explicit?

Part of mentoring is tackling the thorny problem of equalizing opportunities and overcoming prejudice so that 'every pupil is given the opportunity to reach their potential'. Telling is not enough. As mentors we must embody good practice in this.

Setting high but *appropriate* expectations to enable all pupils to attain their potential must be demonstrably clear. To say 'do as I say and not as I do' cannot be right.

Sometimes when teaching report forms are returned to the PGCE office at higher education institutions, sections of the 'other professional requirements' are left blank or a trainee seems to have made little progress within a particular competence.

Commonly neglected areas within Section D of the standards for QTS are 'teachers' professional duties as set out in the current School Teachers' Pay and Conditions Document issued under the School Teachers' Pay and Conditions Act 1991'. What a strange reflection on our profession that trainees can be left unaware of their duties! How should we interpret this omission? Is it that mentors are not sufficiently aware of their duties either, or that they are so implicit in their work that the trainee is unable to decipher what these duties amount to? Knowing one's duties is paramount!

Another requirement is that 'those to be awarded QTS should, when assessed, demonstrate that they . . . keep up to date with research . . . in the subjects they teach.'

On the one hand we demand that scientists tell us what is and what isn't safe for us to eat, yet on the other some teachers persist in rejecting or ignoring how good-quality research could improve their practice in the classroom. Perhaps it has been the language of research that puts many teachers off. Increasingly, research is moved into the classroom and is no longer the territory of the academic working in a university or some distant laboratory. Research is alive and thriving. As a mentor you need to be up to date with the latest developments in pedagogy and research that will enable you to be a better teacher too. It is time for the myth, that teachers in schools train novices in practice and that HE colleagues in HE train them in theory and *research*, to be discarded. It is essential to build a school into a learning community and embedding research into everyday teaching activity.

## Illustration

At X school the deputy head has established a group of action researchers among the staff. There is a thriving atmosphere in the staffroom where several teachers are intent on improving their own practice in the classroom. They work together, observing one another's lessons, offering support and advice, and together they are amassing a body of evidence-based research to draw on as they decide on school practices and new school policies. As an added incentive they have enrolled on a course of study at the local HEI with whom they are working in partnership to seek accreditation for their research at Masters level. What an inspiration for trainees!

## Commentary

This group is firmly taking responsibility for its own professional development. In Chapter 4 you can read about how to carry out research – and the same principles apply to undertaking enquiries in mentoring as they do for classroom teaching.

Increasingly, teachers are being expected to take responsibility for their own professional development. There is ample support but the means of accessing it is changing. The days of spending time out of school on INSET are drawing to a close. Of course, there are still excellent courses available and where funding allows these can be well worth the effort of attending, but INSET is increasingly available on line.

## Accessing sources for continuing professional development

If you access the Internet site for the Teacher Training Agency (http://www.teach-tta.gov.uk) there is a gold mine of information in store for you including details of how to access funding to enable you to carry out research into your own practice in your school.

Perhaps you intend to develop your own in-service training in school; there are details on the TTA site. Similarly you can easily find out about using ICT in your teaching. An essential aspect of mentoring (except for ICT specialists and teachers with a strong personal interest in ICT) that, as yet, remains undeveloped is related to engaging your trainees in good practice in using

ICT in their lessons. At present many trainees are driving the process and training their mentors in using the Internet, for example. This situation is set to change with the implementation of the present government's move to provide ICT training for all teachers in schools, but the government-sponsored training will be insufficient in itself to develop mentoring.

It takes time and effort to develop sound ICT skills and sometimes mentors may feel that there are more pressing demands on their time. But are there? Can we really, as a profession, continue to ignore the opportunities that computers afford us in school?

Computers are not just for pupils' learning: they are a means of developing our skills and knowledge and as mentors we are under a responsibility to stress this to trainees.

Let us take a look at just a few of the opportunities we presently have for professional development using the Internet; the mission statement for the Department for Education and Employment reads:

> to give everyone the chance, through education, training and work, to realise their full potential, and thus build an inclusive and fair society and a competitive economy – that's our aim. To see how we are delivering it, click here . . .

So at http://www.dfee.gov.uk this statement makes inviting reading even before we begin to delve into information about 'the learning gateway'.

On the TTA site (http://www.teach-tta.gov.uk) you can find details about the National Standards for our profession. The main aims of the National Standards are to:

- set out clear expectations for teachers at key points in the profession;
- help teachers at different points in the profession to plan and monitor their development, training and performance effectively, and to set clear, relevant objectives for improving their effectiveness;
- ensure that the focus at every point is on improving the achievement of pupils and the quality of their education;
- provide a basis for the professional recognition of teachers' expertise.

Your trainee needs to know more and you can access information at this site!

Perhaps you want to share more information about current standards for pupils? Accessing the standards site at the DfEE address given is a matter of touching a key. Do you need to know more about the National Curriculum? It is there on-line at http://www.nc.uk.net. Are you up to date with the latest about qualifications and assessment in schools? Try the site for the Qualifications and Curriculum Authority at http://www.qca.org.uk where you will find details about Education 3–16, Lifelong Learning, Key Skills and the latest news relating to this Authority, which directly affects your work in school. How can you neglect such an important source of information in your mentoring?

There is a virtual teacher site that you can access from the DfEE address above. Here you can access schemes of work and subject-specific resources for teaching.

Never before in education has the potential for professional development been so great but it is important in your mentoring to stress that responsibility for CPD increasingly lies with individuals now, whereas once it was all provided for schools.

Becoming a teacher means developing a high degree of ICT awareness and so does effective mentoring. It is likely that mentoring will become supported increasingly by Web-based resources and you can already access mentor development and support.

## Conclusion

If we are genuinely committed to pursuing improvement in our standing as a body of professionals we must effectively create our own standards of 'professionalism' through our work with our pupils, with our colleagues and with entrants to teaching.

We create and hold within our profession deep knowledge about what it is to teach. Through the process of mentoring we can create new knowledge as we work with our trainee teachers. Mentoring is not something we do to mentees, it is something we do with them. We learn as they learn, we educate them as they educate us. We share.

The future of our profession is in our hands. It is no use complaining if standards are imposed on us as professional educators and do not represent what they should. We mentors are gatekeepers to teaching and we create those standards through mentoring.

You can see how mentors create these professional standards within their own practice at my home page, mentorresearch.net. Do let me know about your own experiences of mentoring too!

Questions to ask yourself:

- What do you *know* makes a 'good' teacher?
- How can you work with your trainee so you both become 'good' teachers?

Good practice in being a professional educator:

- Model good practice by your own professional demeanour in school.
- Use the Internet for information about legislation for your trainee.
- Make clear the implications of legislation within your own teaching.
- Model how to take responsibility for our own professional development.

# Glossary

| | |
|---|---|
| CPD | Continuing professional development |
| GTTR | Graduate Teacher Training Registry |
| HE | Higher education |
| HEI | Higher education institution |
| HoD | Head of department |
| ICT | Information and communications technology |
| ITE | Initial teacher education |
| ITT | Initial teacher training |
| LEA | Local education authority |
| OFSTED | Office for Standards in Education |
| PGCE | Post Graduate Certificate in Education |
| QTS | Qualified teacher status |
| RE | Religious Education |
| SCITT | School-centred initial teacher training |
| SENCO | Special educational needs co-ordinator |
| SMT | Senior management team |
| SoW | Scheme of work |
| TTA | Teacher training agency |

# References

Anderson, EM and Shannon, AL (1988) Towards a conceptualisation of mentoring, *Journal of Teacher Education*, **39** (1) 38–42.

Barker S *et al* (1996) *Initial Teacher Education in Secondary Schools,* Teacher Development Research and Dissemination Unit, Institute of Education, University of Warwick, Warwick.

Berrill, M (1994) ITE Cross-roads or by-pass? *Cambridge Journal of Education,* **24** (1), pp 113–15.

Berrill, M (1997) The truth about SCITT. *Managing Schools Today*, March, pp 32–33.

Best, R, Lang, P, Lodge, C and Watkins, C (1995) *Pastoral Care and Personal Social Education: Entitlement and Provision*, Cassell, London.

Brooks, V and Sikes, P (1997) *The Good Mentor Guide: Initial teacher education in secondary schools*, Open University Press, Buckingham.

Brown, JD (1998) *The Self* , McGraw Hill, Boston USA.

Buber M (1947) *Between Man and Man,* Collins, Fontana Press, London.

Button L (1982) *Group Tutoring for the Form Tutor,* Hodder & Stoughton, London.

Calderhead, J and Elliott, B (1993) Mentoring for teacher development: possibilities and caveats, in *Issues in Mentoring,* ed T Kerry and A Shelton Mayes, Routledge/OUP, London.

Calderhead, J and Gates, P (1993) *Conceptualising Reflection in Teacher Development,* Falmer Press, London.

Calderhead, J and Shorrock, S (1997) *Understanding Teacher Education,* The Falmer Press, London.

Capel, S, Leisk, M and Turner, T (1995) *Learning to Teach in the Secondary School,* Routledge, London.

Daloz, A (1986) *Effective Teaching and Mentoring,* Jossey-Bass, San Francisco.

DeBolt, G (ed) (1992) *Teacher Induction and Mentoring: School-Based Collaborative Programmes,* State of New York Press, New York.

Department of Education and Science (1967) The Plowden Report: Children and their primary schools, Department of Education and Science, London.

De Vries, P (1996) *The Management of School Consortia involved in School-centred Initial Teacher Training: A Case Study*, A Report to the Open University Validation Services, London.

Feiman-Nemser, S (1998) Teachers as Teacher Educators (1), *European Journal of Teacher Education*, **21** (1), pp 63–71.

Fletcher, S (1997a) From Mentor to Mentored, *Mentoring and Tutoring*, Summer 1997.

Fletcher, S. (1997b) ITE and Form Tutoring: A question of Responsibility, *Mentoring and Tutoring*, January 1997, pp 45–51.

Fletcher, S (1997c) *Modelling Reflective Practice for Pre-Service Teachers: The Role of Teacher Educators*. Falmer Press, London.

Fletcher, S (1999) Partnerhip in School-based Teacher Education, report prepared in collaboration with the Standing Committee of Partnershp Administrators (unpublished).

Fletcher, S and Calvert, M (1994) *Working with Your Student Teacher*, Mary Glasgow-Stanley Thornes/ALL, Cheltenham.

Fromm, E (197) *Escape From Freedom*, Avon Books, New York.

Fukuyama, F (1992) *The End of History and the Last Man*, Penguin Press, London.

Furlong, J and Maynard, T (1995) *Mentoring Student Teachers*, London: Routledge.

Furlong J and Smith, R (eds) (1996) *The Role of Higher Education in Initial Teacher Training*, Kogan Page, London.

Gagné and White (1978) Memory structures and learning outcomes, *Review of Educational Research*, **48**, pp 187–222.

Gay, B and Stephenson, J (1998) The mentoring dilemma: guidance and/or direction? *Mentoring and Tutoring*, **6** (1/2), pp. 43–54.

Ghaye, A and Ghaye, K (1998) *Teaching and Learning through Critical Reflective Practice*, David Fulton Publishers, London.

Griffiths P and Sherman K (1991) *The Form Tutor: New approaches to tutoring in the 1990s*, Blackwell, Oxford.

Hagger, H, Burn, K and McIntyre, D (1993) *The School Mentor Handbook*, Kogan Page, London.

Hale, R (1999) *To Mix or Mis-match? The Dynamics of Mentoring as a Route to Personal and Organisational Learning*, paper presented at the European Mentoring Conference St. Neots, Cambs, November.

Hamblin, DH (1983) *The Teacher and Counselling*, Blackwell, Oxford.

Hamblin, DH (1989) *A Pastoral Programme*, Blackwell, Oxford.

Hargreaves, A (1998) The Emotional Practice of Teaching, *Teacher and Teacher Education*, **14** (8), pp 835–54.

Harland J *et al* (1999) *Thank You for the Days? How Schools Use Their Non-contact Days*, NFER, Slough.

Hawkey, K (1996) Image and the Pressure to Conform in Learning to Teach, *Teaching and Teacher Education*, **12** (1), pp 99–108.

Hawkey, K (1995) *Peer Support and the Development of Metalearning in School-based Initial Teacher Education*, EDRS availability (University of Bath).
Holly, P and Southworth, G (1989) *The Developing School*, Falmer Press, London.
Lawlor, S (1995) Teachers mistaught, in *An Education Choice: Pamphlets from the Centre 1987–1994*, ed S Lawlor, The Centre for Policy Studies, London.
Marland, M (1974) *Pastoral Care*, Heineman, London.
Maynard, T and Furlong, J (1993) 'Learning to teach and models of mentoring' in *Mentoring Perspectives in School Based Teacher Education*, eds D McIntyre, H Hagger and ME Wilkin, Kogan Page, London.
McIntyre, D and Hagger H (ed) (1996) *Mentors in Schools: developing the profession of teaching*, Fulton, London.
McIntyre, D, Hagger, H and Burn, K (1993) *The School Mentor Handbook: Essential skills and strategies for working with student teachers*, Kogan Page, London.
McIntyre, D, Hagger, H and Burn, K (1994) *The Management of Student Teachers' Learning*, Kogan Page, London.
McNiff, J (1994) *Action Research: Principles and Practice*, Routledge, London.
Mitchell, C and Weber, S (1999) *Reinventing Ourselves as Teachers: Beyond Nostalgia*, Falmer Press, London.
Moyles, J, Suschitzky, W and Chapman, L (1998) *Teaching Fledglings to Fly? Mentoring and Support Systems in Primary Schools*, School of Education, University of Leicester, Leicester.
NAPCE (1991) *Tutor Review* (broadsheet), National Association for Patoral Care in Education, Coventry.
Nicholls, G (1999) *Learning to Teach*, Kogan Page, London.
Nutt, G and Abrahams M (1994) Physical education in teacher education: the Gloucestershire Initial Teacher Education Partnership, *British Journal of Physical Education*, Winter 1994, pp 33–40.
Randell, G, Packard, G and Slater, J (1984) *Staff appraisal: a first step to effective leadership*, Institute of Personnel Management, London.
Reynolds, K (1995) The Role of the Form Tutor, *Pastoral Care in Education*, **13** (3), pp 29–33.
Rowley, J (1999) The good mentor, *Educational Leadership*, **56** (8), 20–22.
Shaw, R (1992) *Teacher Training in Secondary School*, Kogan Page, London.
Schön, D (1987) *Educating the Reflective Practitioner*, Jossey-Bass, San Francisco.
Schön, D (1984) *The Reflective Practitioner: How professionals think in action*, Basic Books, New York.
Shulman, L (1986) Those who understand: knowledge growth in education, *Educational Researcher*, **15**, 4–14.
Simco, N (1999) paper presented at the annual conference for the British Educational Research Association, University of Sussex, Brighton.
Teacher Training Agency (1998) *A National Framework for Professional Development incorporating the National Standards for Qualified Teacher Status, Head Teachers, Subject Leaders and Special Educational Needs Coordinators*, TTA, London.

Tickle, L. (1999) *Teacher Induction: Limits and Possibilities based on Teacher Induction: The Way Ahead*, paper presented at the annual conference for the British Educational Research Association, University of Sussex, Brighton.

Times Educational Supplement (January 1997) *The TES Guide for First Appointments.*

Times Educational Supplement (January 2000) *The TES Guide for Newly Qualified Teachers.*

Tomlinson, P (1995*) Understanding Mentoring: reflective strategies for school-based teacher preparation*, Open University Press, Buckingham.

University of Bath (1999–2000) *Post Graduate Certificate in Eduaction Course Handbook*, University of Bath, Bath.

University of Bath (1999–2000) *Subject Didactics Handbook for Modern Foreign Languages Post Graduate Certificate in Eduaction Course*, University of Bath, Bath.

University of Bath (1999–2000) *Form Tutor Digest, Post Graduate Certificate in Education Course*, University of Bath, Bath.

University of Bath (1999–2000) *Professional Development Plan, Post Graduate Certificate in Education Course*, University of Bath, Bath.

Watkins, C and Whalley, C (1993*)* Mentoring beginner teachers – issues for schools to anticipate and manage, School Organisation, **13** (2), pp 129–38.

Whitehead, J (2000) How do I improve my practice? Creating and legitimating an epistemology of practice, *Reflective Practice,* **1** (1), pp 91-104.

Whitehead, J (1989) Creating a living educational theory from questions of the kind 'How do I improve my practice?' *Cambridge Journal of Education*, **19** (1) pp 42–51.

Wilkin, M (1992) (ed) *Mentoring in schools*, Kogan Page, London.

Wilkin, M and Sankey, D (eds) (1994) *Collaboration and Transition in Initial Teacher Training*, Kogan Page, London.

Wootton, M (1992) *Being a Form Tutor, On Your Own In The Classroom*, no. 14, Upminster: Nightingale Teaching Consultancy, Upminster.

Yamamoto, K (1990) To see life grow: the meaning of mentorship, *Theory into Practice*, **17** (3), pp 183–89.

# Further reading

Adams, S (1989) *A Guide to Creative Tutoring: The Tutor Ascendant*, Kogan Page, London.

Adler, S (1991) The reflective practitioner and the curriculum of teacher education, *Journal of Education for Teaching*, **17** (2), pp 139–50.

Baldwin, J and Wells, H (1979) *Active Tutorial Work*, Blackwell, Oxford.

Bennett, C, Jones, B and Maude, P (1994) Insights from the Teaching Profession: potential advantages and disadvantages of greater school involvement in ITE, *Cambridge Journal of Education*, **24** (1), pp 67–73.

Berrill, M (1993) It's time we were frank, Mr. Berrill, *Times Educational Supplement*, January 8.

Berrill, M (1997) *A Foundation for Excellence: A Structured Approach to Mentoring and the development of basic Teaching Proficiency*, Chiltern Training Group, Bedfordshire (SCITT).

Best, R, Ribbins, P, Jarvis, C and Oddy, D (1993) *Education and Care*, Heinemann, London.

Blake, D and Hill, D (1995) The newly qualified teacher in school, *Research Papers in Education*, **10** (3), pp 309–39

Brighouse, T and Woods, D (1999) *How to Improve your School*, Routledge, London.

Brooks, V, Barker, S and Swatton, P (1997) Quid pro quo? Initial teacher education in secondary schools, *British Educational Research Journal*, **23** (2), pp 163–78.

Cameron-Jones, M (1993) Must a mentor have two sides? *Mentoring and Tutoring*, **1** (1), pp 5–8.

Cameron-Jones, M (1995) Mentors' perceptions of their roles with students in initial teacher training, *Cambridge Journal of Education*, **25** (2), pp 189–99.

Capel, S, Leisk, M and Turner, T (1997) *Starting to Teach in the Secondary School*, Routledge, London.

Cohen, NH (1995) *Mentoring Adult Learners – A guide for educators and trainers*, Krieger Publishing Company, Florida.

Cross, KP (1986) *Adults as Learners,* Jossey-Bass, San Francisco.

De Vries, P (1996) Could criteria used in quality assessments be classified as academic standards? *Higher Education Quarterly,* 50 (3), pp 193–206.

Dunne, R and Harvard, G (1994) A model of teaching and its implications for mentoring, in *Mentoring: Perspectives on School-Based Teacher Education,* ed D McIntyre, H Hagger, and M Wilkin, pp 117–29, Kogan Page, London.

Elliott, J (1993) *Action Research for Educational Change,* Open University Press, Milton Keynes.

Elliott, B and Calderhead, J (1994) Mentoring for teacher development: possibilities and caveats, in *Mentoring: Perspectives on School-based Teacher Education,* ed D McIntyre, H Hagger and K Burn, Kogan Page, London.

Evans, L (1997) The practicalities involved in the introduction of school-administered initial teacher education in the United Kingdom; some policy issues and implications, *Research Papers in Education,* **12** (3), pp 317–38.

Evans, L, Abbott, I, Goodyear, R and Pritchard, A (1996) in *Mentoring and Tutoring,* **4** (1) Summer.

Feiman-Nemser, S (1987) When is student teaching teacher education? *Teaching and Teacher Education,* **3** (4), pp 255–97.

Fidler, B and Lock, N (1994) Mentorship and whole school development, in *Improving Initial Teacher Training?* in M McCulloch and B Fidler, pp 141–155, BEMAS/Longman, Harlow.

Fish, D (1989) *Learning through Practice in Initial Teacher Training,* Kogan Page, London.

Fletcher, S (1995) Caveat mentor, *Language Learning Journal for the Association for Language Learning,* March.

Fletcher, SJ (1998) Attaining self-actualisation through mentoring, in *European Journal of Teacher Education,* **21** (1), pp 109–18.

Furlong, J (1994) Another view from the cross-roads, *Cambridge Journal of Education,* **24** (1), pp 117–20.

Furlong, J, *et al* (1988) *Initial Teacher Training and the Role of the School,* Open University Press, Milton Keynes.

Griffiths, V and Owen, P (1995) *Schools in Partnership,* Paul Chapman, London.

Hagger, H and McIntyre, D (1994) *Mentoring in Initial Teacher Education.*

Hannan, A (1995) *The case for school-led primary teacher training, Journal of Education for Teaching,* **21** (1), pp 25–35.

Harris, A, Jamieson, I and Russ J (1996) *School Effectiveness and School Improvement: A Practical Guide,* Pitman, London.

Hawkey, K (1997) *Roles, Responsibilities and Relationships in Mentoring: A literature review and agenda for research, Journal of Teacher Education,* November–December, **48** (5), 323–36.

Hawkey, K (1998) Consultative supervision and mentor development, *Teachers and Teaching: Theory and Practice,* **4** (2), pp 331–42.

Howie, C (1995) *Can School Mentors Aid the Development of Reflective Practice in Student Teachers*? Paper presented at European Conference for Educational Research at the University of Bath, September 1995.

Hustler, D and McIntyre, D (1996) *Developing Competent Teachers*, David Charles, London.

Hutchinson, D (1994) Competence-based profiles for ITT and induction: the place of reflection, *British Journal of In-Service Education*, **20** (3).

Kramm, K (1983) Phases of a mentor relationship, *Academy of Management Journal*, December.

Kyriacou, C (1997) *Effective Teaching in Schools*, Stanley Thornes, Cheltenham.

Laidlaw, M (1992) *Action Research: A Guide for Use on Initial Teacher Education Programmes*, University of Bath, Bath.

Leithwood, K and Louis, KS (1998) *Organizational Learning in Schools*, Swets and Zeitlinger, Abingdon.

Lortie, D (1994) Teacher career and work rewards, in *Classrooms and Staffrooms: the sociology of teachers and teaching*, ed A Hargreaves and P Woods, Open University Press, Milton Keynes.

McCulloch, M and Lock, N (1994) Mentorship developments in the primary phase of initial teacher education at the University of Reading, *Mentoring*, **1** (3), pp 21–28.

Miles, S, Everton, T and Bonnett, M (1994) Primary Partnership Matters: some views from the profession, *Cambridge Journal of Education*, 24 (1), pp 49–65.

Mullen, C and Lick, D (ed) (1999) *New Directions in Mentoring*, Falmer Press, London.

Nathan, M (1995) *The New Teacher's Survival Guide*, Kogan Page, London.

Norman, N (1995) Initial teacher education in France, Germany and England and Wales: a comparative perspective, *Compare*, 25 (3), pp 211–26.

Parsloe, E (1995) *Coaching, Mentoring and Assessing: A practical guide to developing competence*, Kogan Page, London.

Parsloe, E and Wray, M (2000) *Coaching and Mentoring: Practical Methods to Improve Learning*, Kogan Page, London.

Roth, RA (ed) (1998) *The Role of the University in the Preparation of Teachers*, Falmer Press, London.

Russell, T (1988) From pre-service teacher education to first year of teaching: a study of theory and practice, in *Teachers' Professional Learning*, (ed) J Calderhead, Falmer Press, London, pp 13–34.

Russell, T (1997) Teaching teachers: how I teach is the message, in *Teaching about Teaching: Purpose, Passion and Pedagogy*, (ed) J Loughran and T Russell, Falmer Press, London.

Sikes, P (1992) The Life Cycle of the Teacher, in *Teachers Lives and Careers*, (ed) S Ball and I Goodson, Falmer Press, London, pp. 27–60.

Simco, N and Sixsmith, SC (1994) Developing a mentoring scheme in primary initial teacher education, *Mentoring and Tutoring*, **2** (1), pp 27–31.

Stern, J (1995) *Learning to Teach*, David Fulton Publishers, London.
Tilley, S (1994) The role of the work based mentor, *Training Officer,* **30** (3), p 84.
Trend, R (1998) *Qualified Teacher Status: A practical introduction*, Letts Educational, London.
Vlaeminke, M (1996) *The Active Mentoring Programme: Pack 3, developing key subject competences*, Pearson Publishing, Cambridge.
Wajnryb, R (1993) *Classroom Observation Tasks*, Cambridge University Press, Cambridge.
Wallace, M (1991) *Training Foreign Language Teachers: a reflective approach*, Cambridge University Press, Cambridge.
Waterhouse, P (1991) *Tutoring*, Network Educational Press, Stafford.
Watkins, C and Thacker, J (1993) *Tutoring INSET Resources for a Whole School Approach*, Longman, Harlow.
Watkins, C and Whalley C (1993) *Mentoring Resources for School-based Development*, Longman, Harlow.
Whiting, C *et al* (1996) *Modes of Teacher Education Project: Partnerships in Initial Teacher Education: A Topography*, Institute of Education, London.
Wilkin, M (1992) The challenge of diversity, *Cambridge Journal of Education*, **22** (3), pp 307–22.
Wilkin, M *et al* (1997) *The Subject Mentor Handbook for the Secondary School*, Kogan Page, London.
Williams, A (1994) *Perspectives on Partnership,* Falmer Press, London.
Williams, A (1995) Supporting student teachers in sc hool-based work: some policy implications, *Educational Management and Administration*, **23** (3), pp 188–96.
Zeichner, K and Liston, D (1987) Teaching student teachers to reflect, *Harvard Educational Review*, **57** (1), 23–45.

# Index